Hitch Your Antenna to the Stars

Hitch Your Antenna to the Stars

Early Television and Broadcast Stardom

SUSAN MURRAY

Routledge
Taylor & Francis Group

NEW YORK AND LONDON

Versions of Chapter 3 have appeared as "Ethnic Masculinity and Early Television's Vaudeo Star," in *Cinema Journal* (Austin: University of Texas Press) 42, no. 1, Winter 2002: 97–119 and as "Lessons from Uncle Miltie: Ethnic Masculinity and Early Television's Vaudeo Star," in *Small Screens, Big Ideas: Television in the 1950s*, ed. Janet Thumim (London: I. B. Tauris) 2002: 66–87.

A version of Chapter 5 has appeared as "Our Man Godfrey: Arthur Godfrey and the Selling of Stardom in Early Television," in *Television and New Media* (Thousand Oaks, CA: Sage Publications) August 2001: 187–203.

Published in 2005 by
Routledge
Taylor & Francis Group
270 Madison Avenue
New York, NY 10016

Published in Great Britain by
Routledge
Taylor & Francis Group
2 Park Square
Milton Park, Abingdon
Oxon OX14 4RN

© 2005 by Taylor & Francis Group, LLC
Routledge is an imprint of Taylor & Francis Group

Printed in the United States of America on acid-free paper
10 9 8 7 6 5 4 3 2 1

International Standard Book Number-10: 0-415-97130-6 (Hardcover) 0-415-97131-4 (Softcover)
International Standard Book Number-13: 978-0-415-97130-0 (Hardcover) 978-0-415-97131-7 (Softcover)
Library of Congress Card Number 2005006737

Library of Congress Cataloging-in-Publication Data

Murray, Susan, 1967-
   Hitch your antenna to the stars! : early television and broadcast stardom / Susan Murray.
     p. cm.
   ISBN 0-415-97130-6 (hardback : alk. paper) -- ISBN 0-415-97131-4 (pbk. : alk. paper)
    1. Television broadcasting--United States--History. 2. Television actors and actresses--United States--Biography. 3. Television programs--United States--History. I. Title.

PN1992.3.U5M87 2005
791.45'0973--dc22
                                             2005006737

Taylor & Francis Group
is the Academic Division of T&F Informa plc.

Visit the Taylor & Francis Web site at
http://www.taylorandfrancis.com

and the Routledge Web site at
http://www.routledge-ny.com

CONTENTS

ACKNOWLEDGMENTS

There were many people and institutions that helped bring this book to fruition. My dissertation committee at the University of Texas at Austin helped shape earlier versions of this project and pointed me in the right direction for further research. Thomas Schatz, Mary Desjardins, Horace Newcomb, Rosa Eberly, and Hilary Radner were consistently generous with their time and insight. Janet Staiger, my advisor, was meticulous in her comments and challenged me in such a way as to make me a more thorough researcher and a better thinker overall. I also thank William Boddy, Justin Wyatt, and Janet Thumim for reading parts or all of this manuscript during its transition from a dissertation to a book. A shared interest in Arthur Godfrey led to many enlightening conversations and e-mail exchanges with Douglas Gomery. Ever since, he has championed my work and become a cherished friend.

The librarians and archivists at the Wisconsin Historical Society, the Motion Picture, Broadcasting and Sound Division of the Library of Congress, and the UCLA Film and Television Archive guided me to vital material. Special thanks go to Michael Henry at the Library of American Broadcasting at the University of Maryland College Park for his expertise, patience, and generosity. While researching and writing this work, I was supported by a William T. Livingstone dissertation fellowship at the University of Texas at Austin, a Goddard Award from the Steinhardt School at New York University, and an American Post-doctoral Fellowship from the American Association of University Women.

I also thank Matthew J. Byrnie at Routledge for his patience, efficiency, and enthusiasm for the project; Andrew Hearst and Vanessa Grimm for their expert copyediting and fact checking skills; and Noah Shibley for his help with the cover design.

My colleagues in the department of Culture and Communication and in the broader community of New York University have provided a rich and supportive working environment. I am particularly grateful for the encouragement and friendship of Aurora Wallace, Ted Magder, Alex Galloway, Neil Postman, Christine Nystrom, Andrew Light, Mitchell Stevens, Anna McCarthy, and Toby Miller.

There were also a number of friends who lent their support, patience, company, expertise, and, at times, their guest rooms and couches while I was working on this project. For this, I would like to thank Amy Bowles-Reyer, Lauren Bowles, Jonathan Burston, Michael DeAngelis, Erich Dietrich, Patrick Fischler, Vanessa Grimm, Heather Hendershot, Arvind Kannabiran, Christina Lane, Alison Macor, Anne Morey, Laurie Ouellette, M. J. Robinson, Kristin Ruble, Cathy Simmons, Irene Sosa, Malcolm Turvey, and Federico Windhausen.

Finally, I have two families of Murrays to thank. First of all, I am indebted to my parents, Donald and Dorrit, and my brother, Steve, for everything they have given to me over the years. In terms of this project, I will always treasure the image of my father sitting down in his living room to read my dissertation from start to finish, even though I know that his true interest in the subject was limited. Although the other family of Murrays—Ben, Joyce, Liam, and Siobhan—are not biologically related to me, it has often felt as though they were. The many nights I shared with them sitting around their kitchen table in Australia, long after our plates were cleared, drinking wine and discussing the issues of the day, solidified my love for ideas and debate.

Introduction

A 1955 print ad shows television fixture Jackie Gleason being upstaged by clothing from his signature "Jackie Gleason Originals" line by Manhattan Shirt Company. The ad plays off of Gleason's reputation as a showman—a persona he reminds us of by hamming it up in a Santa suit with his eyes bulging and hands pressed against his chest in an exaggerated expression of surprise—by suggesting that the star plays second fiddle to the very products he's endorsing. Gleason looks on with astonishment (or perhaps intrigue, or envy) as the CBS television camera, microphone, and arc lamp are directed not at him but instead at a selection of dress shirts, ties, pajamas, and handkerchiefs—some of which bear illustrated figures of Gleason in his signature "And away we go!" stance—that are placed front and center in the ad, shunting him off to the side.

In some ways, this ad is nothing unusual. Magazines from the postwar era were littered with ads featuring television stars hawking merchandise related to their on-screen personas in one way or another. However, the image of Gleason being displaced by his own products is worth noting because it speaks directly to a significant shift taking place in the television industry right around the time of the ad's creation. By 1955 the industry was moving out of its transitional early years and was in the process of standardizing not only many of its production and programming practices but also the forms and functions of its star system. By this point, the selling and the merchandising of one's own persona had become an absolutely essential component of a television star's career.

Fig. I.A Jackie Gleason sells his own collection of menswear for Manhattan Shirt Company, 1955.

Much like the clothing in the ad, the star branded or promoted product was taking center stage in television's economy.

A number of interlocking interests created and circulated through a television star's persona during this period. For instance, by appearing in the Manhattan Shirt ad, Gleason was simultaneously serving the interests of the advertiser, his network, an advertising agency, his talent agency, and his own company, Jackie Gleason Enterprises. The figures on the clothing and the hyperbolic gestures of his still image in the ad stimulated viewer identification of Gleason's grandiose performance style and persona. Like the majority of variety stars who dominated prime-time postwar television, Gleason was a highly intertextual performer. He worked in a range of entertainment sites prior to his entry into television, and once he became a star he lent his image to numerous commercial endeavors. His intertextuality and commercial malleability helped make him famous with postwar consumers. Constructed by his television programs, press accounts, public appearances, endorsements, and larger discourses about television and stardom, his image moves through interpenetrating commercial and entertainment narratives with the aim of garnering viewers and stimulating sales.

Television stars were the most visible manifestation of the operations and aims of the broadcast industry during the postwar period. They were used to sell or promote television sets, programs, and consumer goods, and represent network and social identities. Many of television's early comedy stars also were used to reference the already familiar entertainment forms of vaudeville, radio, and film as they helped delineate television's distinctive aesthetic and narrative characteristics. Such competing roles and responsibilities were foundational qualities of the highly regulated, yet decentralized, process of broadcast star production and management; one that was notably different from the Hollywood star system of the studio era. For example, during the early 1950s, as the broadcast industry was channeling profits from radio into its new medium of television, the interest and investment into programs and performers was spread among networks, sponsors, talent agencies, and advertising agencies. A typical network-star contract regulated the performer's guest appearances on other shows and specifically stipulated that the performer, or his or her name, could not appear on any other network. Many times, performers were not only contracted to a network but also with the sponsor of their program, giving the company full and exclusive use of the performer's likeness for the purposes of advertising. Instead of conforming to the more unified aims of a single studio, which only sold movies and related merchandise, the television star was required to advertise a product while also representing the textual and industrial strategies of a television network. Hence, television on-screen talent had to represent the sometimes competing commercial aims of both the sponsor (who wanted the star to be associated with its specific product) and the network (who used the star's persona to represent the character of the network as well to attract a mass audience). The star was utilized as a part of product differentiation, but for at least two very different products, which meant that the commercialism of a performer's persona in television was much more overt and the system in which it functioned was more diffuse than it had been in Hollywood.

Such strategies of star cultivation and management were developed in relation to radio and in regards to the industrial and cultural understanding of what television was on its way to becoming. Although long anticipated by both consumers and retailers, television sets were not readily available until the summer of 1946 and even then were out of reach pricewise for the average American consumer. Moreover, even if you purchased a set at that time, there was a relatively limited selection of programs on the air and a very spotty distribution of licensed television stations, most of which were located in major cities. Despite their limited audience, the networks (NBC and CBS in particular) were experimenting with programming formats and production practices as well as sorting out what type of performer would be best suited to the medium's specific aesthetic qualities and would attract the largest number of viewers and advertisers.

One of the main issues surrounding television and its development of a star system was its tendency to poach talent from other entertainment industries instead of working to cultivate its own. In January 1953, *Los Angeles Mirror* radio-TV editor Hal Humphrey dedicated his weekly column to the topic of television stardom. "For some reason TV seems to lack the ability to establish its own big names or star personalities," he complained.[1] "Strip away all the hoopla about the medium's tremendous impact on the entertainment world and you'll find that the only stars which TV actually can call its own are Gorgeous George, Baron Leone, Howdy Doody, Beany and Kukla, Fran and Ollie."[2] By insisting that a television star must be a performer who *initially and solely* found his/her fame on television, Humphrey concluded that only puppets and wrestlers fit the bill. As early as the late 1940s, critics were expressing similar disappointment with what appeared to be television's inability to be a "star-maker." Many of them had hoped the coming of a new broadcast medium would immediately lead to the production of new faces and, in particular, the creation of glamorous dramatic stars. What they had, they complained, was the same old formats and aging comedians they had heard on radio and seen on the vaudeville circuit.

Historical debates such as these are certainly intrinsic to the meaning and function of broadcast stardom, but they also are essential to understanding the development of television and the relationships it fostered with other entertainment fields such as vaudeville, radio, Broadway, legitimate theater, and film. Beyond the synergistic promotional and investment relationships that television initiated with these industries, its use of stars during the postwar period revealed the way it borrowed, referenced, and recast certain performance styles, genres, and cultural traditions as it simultaneously attempted to distinguish itself as a unique and innovative medium. Television used its stars to define itself.

Although this book is clearly about the industrial and cultural work performed by broadcast stars during this period, it is also a book about comedy. More than any other type of performer, comedians were the ones that most fully embodied the industry's commercial and entertainment aims and ambitions. Consequently, the focus of this book is on early television's national prime-time comedy stars: those working in vaudeo and the very first stars of sitcoms appearing on the air between 1947 and the late 1950s; the performers who were the most visible and oft-discussed generic manifestations of the television star during its early years. The vaudeo (video vaudeville) genre, in particular, was the center of much industry expectation. Based on my research, debates about the definitional contours of television stardom and star salaries, issues of the economic and cultural influence of stardom, the development of national star-based advertising, and network promotional campaigns centered on the stars of vaudeo programs. Such debates and issues exposed the ways in which different areas of the broadcast industry imagined and reimagined the possible meanings and implications of stardom for television. A general conception of the vaudeo star acted as the axis on which many of these struggles over meaning occurred.

For the purposes of this project, I have contextualized the negotiation of television stardom in comedy within three primary stages of programming development occurring in the years between the initial appearance of regularly scheduled network programming and the standardization of production and narrative practices in the mid 1950s. These stages are: (1) the recasting of many of radio's advertising and programming practices for television; (2) the television industry's poaching of ex-vaudevillians from stage, film, and radio for the purposes of creating presold stars for the new medium; (3) and, finally, the rise of the telefilm, the domestic sitcom and its stars in the mid-1950s.[3] Within these periods, a number of intensely popular forms of stardom expressed alterations in consumer habits, anxieties over the changing cultural milieu, and shifts in the relationships among major entertainment industries.

The year 1948, in particular, was a time of both growth and circumscription for television: a number of successful variety shows were launched, including two smash hits, *Texaco Star Theatre* and *Toast of the Town*; CBS raided NBC's talent pool by offering radio stars such as

Freeman Godsen, Charles Correll, Edgar Bergen, and Jack Benny deals that included a capital-gains tax rate; and the FCC initiated its "freeze" on television licenses in an effort to contain interference problems caused by the rapid proliferation of new stations. This also was an important year for television stardom, as it was the moment that the debates over what look, experience, and talent a performer may need for television were really heating up, and it marks the inception of a number of the early network star management strategies.

I continue studying the industrial and cultural shifts of the innovation and diffusion period through the late 1950s when filmed episodic series were on the rise. This decade witnessed compelling industrial, generic, economic, and aesthetic shifts in the broadcast industry. It was a time when the industry was testing and developing strategies that would work for the new medium. Eventually, as the industry realized that many of the management, marketing, and economic systems it had used for radio were not going to work for television, the relationship between networks and stars was altered in significant ways. This was in large part because star production and management were central to the eventual standardization of production and narrative modes that developed in the mid-1950s with the rise of independent production and the decline of sponsor control.

The first stage of television's negotiation of stardom is discussed through the history of prime-time network radio and its complex relationships to commercialism, technology, and traditional forms of entertainment. To elucidate the sources of broadcasting's earliest conception of stardom, Chapter 1 focuses on radio's use of top-name talent. The economic structures and narrative strategies honed by radio during its Golden Age were central to the formation of stardom in early television. In this chapter, I trace the cultivation and management of talent in radio from the 1920s until the late 1940s, reconstructing the relationships between networks, local stations, ad agencies, sponsors, and performers, and tracking the movement from indirect advertising (a period wherein acts carried the name of the sponsor) to the acquisition of top-name, pre-sold talent. It ends with an exploration of Eddie Cantor's radio career, as well as a discussion of CBS's infamous 1948 talent raid—and what the raid revealed about the network's real and projected reliance on stars.

Chapter 2 explores how the variety star came to represent the ideal television performer during commercial television's first few years. Because, in large part, of the vaudeo stars' ability to represent what the industry believed were its primary aesthetic properties—immediacy, intimacy, and spontaneity—performers such as Milton Berle became the most popular performers in prime time. In addition, these stars' personas and prior careers in vaudeville, film, and radio were familiar and comforting to the industry's predominantly urban audience. This chapter also will reveal the ways in which radio stars and the rest of the broadcast industry responded to the coming of television. Many stars were initially wary of the new medium, believing that their personas and talent would not transfer well into a medium that emphasized the visual over the aural.

Chapter 3 continues the discussion of vaudeo stars, but, instead of focusing on their relationship to the aesthetics of television and the critical debates that surrounded their popularity, it concentrates on the cultural resonance of their representation of a particular type of ethnic masculinity. This chapter asserts that, although a variety performer's specific ethnic identity was rarely addressed outright in television, traces of his "Jewishness" were embedded in the star's persona and performance style. Specifically, the cultural and religious heritage of individuals such as Milton Berle, George Burns, Sid Caesar, and Jack Benny were obliquely referenced through their connection to the traditions of vaudeville, their affiliations with particular geographic areas/neighborhoods, their relationships with their extended families, and their representations of a historically and culturally located feminized masculinity. The subtle nature of these ethnic cues helped these comedians address the cultural experiences of the largely northern and urban television audience of the late 1940s without completely alienating viewers from other regions.

In Chapter 4, the vaudeo star's decline and the industry's subsequent "talent crisis" in the early 1950s is analyzed in relation to shifting audience demographics, the medium's voracious appetite for material, the increasing power of talent over programming, and the skyrocketing costs of talent and production. The "talent crisis" resulted in the creation of new venues and programs to develop new television talent as well as a rewriting of the variety format to include more situational storylines. In addition, the issues raised by the crisis were a contributing factor in the eventual rise of the sitcom and the decline of the single-sponsor program.

Arthur Godfrey was one of the most popular, prolific, and well-paid stars of the postwar era. In fact, it was said that he alone was responsible for 12 percent of CBS's overall revenues. Chapter 5 uses the construction of Godfrey's persona and its virtual dissolution in 1953 as a way to talk about the role that the notion of authenticity played in both the ability of a star to successfully pitch product as well as its centrality to postwar television's delineation of stardom in general. Chapter 6 focuses on a new paradigm for television stardom that developed in the early to mid 1950s. With the larger context of shifting relationships among sponsors, networks, and independent producers as a backdrop, I discuss the ramifications of the sitcom format and the rise of telefilm on the construction of television stardom, although acknowledging the continuing, albeit altered, presence of the vaudeo style in this process.

This book represents the first major history of early television stardom. Although there are quite a number of well-written star biographies available that reveal the experiences, memories, and opinions of certain individual stars and some great academic histories of the early television industry, there is still no concrete, commonly understood understanding of the initial industrial conceptualization of television stardom and its subsequent permutations. The dearth of information on and analysis of this subject is certainly *not* because stars are an impalpable presence in primary documentation and accounts of postwar broadcasting. On the contrary, stars make up the majority of popular press coverage of television during this period. In addition, the trade press is replete with articles on the long-awaited star system; the careers of individual stars; the role of stars in the networks' processes of standardization and differentiation; and the conflicts surrounding the unionization of television performers. Indeed, it is downright odd that television historians have not yet fully utilized these materials and followed the example of their counterparts in cinema studies who have traced the use and implications of the film industry's star system or those who have used specific stars as case studies for cultural or industrial analyses of discrete historical periods. With the publication of this book, I hope to stimulate more research in this area, as I could only cover a single decade of television's production of image and celebrity. I would like to see others pick up where I left off and explore how the various industrial and commercial relationships, performance styles, personas, and business strategies that helped shape the star system in the postwar era were altered, discarded, or adapted to the 1970s, 1980s, and 1990s.

CHAPTER 1
RADIO AND THE SALIENCY OF A BROADCAST STAR SYSTEM

One cannot fully understand the workings of early television without first understanding network radio. This is true not only because they were part of the same industry, were technologically linked, and had almost identical patterns of funding and ownership, but also because early television borrowed many of radio's cultural signifiers and narrative strategies. In many instances, entire programs, genres, and scheduling techniques were simply lifted from radio and reworked to fit the aesthetic qualities of television. As such, by the 1950 season, many popular variety shows and sitcoms were being simulcast on both mediums during prime time; by mid-decade, soap operas had begun to migrate to television as well.[1] It was commonly thought that such familiar programming would help ease the audience's transition from one broadcasting medium to another. As is the case with most presold products, it was also believed that adapting successful radio shows for television would lower the financial risk because brand new television shows were expensive to produce and carried with them no guarantees of success. Highly popular radio stars were thought to ease this transition as well; audiences would surely be eager to finally *watch* the performers they had spent so many years listening to. However, even as sponsors and networks enjoyed the predictability of presold products, there were concerns that radio

1

programming had already gone stale, and that once the novelty of television wore off viewers would tire of already familiar programs and stars. Many in the industry believed that television would eventually have no choice but to produce unique genres and new faces. As *Variety* reported in March 1948, "With constant criticism raised against radio for its failure to develop new stars or new programming formats there's no question about it being in a static condition. Top network and ad agency personnel, consequently, are hopefully eyeing tele as the prescription that can remedy the situation."[2]

Obviously, television performance and its accompanying star system did not occur in a vacuum. They developed in direct relationship to radio and, to a lesser extent, to film, nightclubs, and theater. Television narratives, performance, and its production of celebrity in some ways had to make up for a perceived lack in radio, while simultaneously maintaining many of its traditions and borrowing on its successes. The industrial and economic structures that supported radio during its Golden Age carried over to television, affecting not only the manner in which stars were utilized, packaged, and sold but also what types of stars would emerge out of and flourish in this environment. For example, as we will see in succeeding chapters, vaudeo star came about logically because of the industry's desire to differentiate itself aesthetically from other mediums, while still balancing the financial interests and needs of the networks, sponsors, advertising agencies, and audiences, most of whom were all familiar and comfortable with what they had experienced in radio. Radio set the terms for the conception of the medium of television as well as furnishing the framework for the development of broadcasting's star system. This chapter will explore the ways that the broadcast industry prefigured the characteristics of the ideal television performer through its generation of radio talent, the interlocking economic relationships it established in its earlier years, and the particular manner in which it negotiated the coupling of entertainment and commercial narratives. In order to unearth the complex relationships that existed among radio, television, film, and stage, we need to turn to the 1920s, a period in which radio went from being a highly localized hobby for amateurs to a form of mass entertainment engaged with national narratives and shifting notions of consumerism and what it meant to be American.

Indirect Advertising and Radio Talent in the 1920s

During the decade following the end of World War I, the United States underwent massive social change. As most are aware, this was the time of prohibition, flappers, modernism, and the Jazz Age. Yet, it was also a time when the liberal values of urban life were coming into conflict with those of rural protestants; when nativists organized in response to pluralism and increasing ethnic heterogeneity; when a simultaneous suspicion of and expansion of federal power existed; and when mass production, urbanization, and popular culture were on the rise. American culture was also becoming more and more of a consumer culture as the growth of advertising and mass media industries coincided with an increase in leisure time and disposable income. According to the social historian Lynn Dumenil, both the working and middle classes "found in the consumer culture an antidote to the loss of power in the modern world, to the problem of hierarchy, routinization, and standardization."[3]

For broadcasting, the 1920s was the period in which it became a centralized, regulated, national industry and a primary producer of cultural texts. By 1927 NBC had two networks on the air (NBC-Red, NBC-Blue) and CBS had just made its national debut. Although radio was well on its way to becoming a popular domestic middle-class entertainment medium by the early to mid-1920s, it did not produce much innovative programming. At first it was primarily unknown singers, commentators, and announcers who filled the airwaves, because, as an avenue for primarily music and information, radio made only small and occasional forays into the production of narrative.[4] Programs featuring orchestras such as the A&P Gypsies and the Ipana Troubadours were the most popular shows on the air for a time.[5] Comedy programs, which would eventually prove to be broadcasting's forte, were not a staple of network radio until the 1930s. In the 1920s, local stations began to hire the occasional low-budget comic for "song and patter" programs but, for the most part, big-name comedians stayed away from the new medium.[6] In his 1979 book, *Radio Comedy*, Frank Wertheim argues that vaudeville and nightclub performers initially were reluctant to perform on radio because they found performing with a microphone without an audience alienating; they weren't willing to accept the low pay that broadcasters were offering; and vaudeville booking agents were reluctant to allow their players to perform in an entertainment medium that was competing with their own.[7] This would all change as broadcasting proved itself to be a true competitor in the world of entertainment and as vaudeville began to die off in the early 1930s. In the meantime, mostly announcers, musical groups, and other types of amateur performers put in their time on local stations. Along the way, many of them became known not by their own names but by the names of sponsors.

The practice of indirect advertising, which involved connecting a sponsor or product name with the program title or performers, was common during the early years of broadcasting primarily because of the debates that had occurred around the regulation and commercialization of the broadcast industry. In a series of conferences occurring from 1922 to 1925 the Department of Commerce (led by Herbert H. Hoover) took on the question of just how radio would be funded and regulated. There was little doubt that they would come down in favor of an open market rather than pure governmental support and operation. Yet there were concerns expressed both in the conferences as well as by the general public regarding the effect that product advertising and private ownership would have on the types of programming produced. In an effort to create a brake of sorts for rampant commercialism and decisions influenced primarily by the bottom line, the legislative proposal that resulted from the conferences included calls for broadcasting to serve the public good. (It should be noted that it was also used to rationalize the support of a few large corporations in their broadcasting endeavors.[8]) The policies that resulted would require this private industry to be held somewhat accountable for its duty to serve the "public interest"—a troublingly vague term that could be used to either protect or justify a range of policies and actions. As Thomas Streeter has pointed out, even broadcast advertising itself was "justified on the grounds that it served the needs of the system, and thus the public interest."[9] Nevertheless, broadcasters were aware of listeners' distaste for brash commercialism and were concerned about being too overt in their commercial objectives, so indirect advertising provided a comfortable middle ground for a time wherein the sponsor's name would be heard but without a hard sell or direct pitch. Edgar H. Felix, whose book *Using Radio in Sales Promotion* was published in 1927, reminded his readers that:

> The method used to direct attention to the sponsor through which the goodwill gained is capitalized by him is the most delicate phase of program preparation. In this respect more than in any other must the attitude of the listener be kept scrupulously in mind. The most successful are those in which the name of the feature itself is indelibly tied in with the name of the product. Eveready Hour, the Happiness Boys, the Gold Dust Twins, Clicquot Club Eskimos, for example, are household words in the areas which are served by their programs. [10]

The Happiness Boys were in fact one of the first acts to appear regularly on the broadcast schedule under their sponsor's name. The Happiness

Candy Company first aired their "song and humor" program featuring the duo Billy Jones and Ernie Hare on WEAF in New York in 1923. The pair had worked in vaudeville and recording prior to being hired for radio. On radio, however, their own names were subordinated to the larger aims of the sponsor.[11] The success of this team seemed to prove the effectiveness of indirect advertising as a negotiative tactic, because as long as Happiness Candy avoided direct advertising, the popularity of their show and the visibility of their product increased. Felix, who was writing during the peak years of indirect advertising, noted, "The 'Happiness Boys' are radio's outstanding humorists, and their following is as loyal as that of any movie star or baseball idol."[12] By late 1928, the Happiness Boys had become the Interwoven Pair for the Interwoven Sock Company. Later, they would become the Taystee Loafers for Taystee, and in 1932 they performed as the Best Food Boys for Best Foods. Although some listeners knew that the Happiness Boys/the Interwoven Pair were actually Jones and Hare, most performers on the radio in the 1920s were anonymous. The only name associated with their performance/persona was the sponsor's. Most of these people were new to the entertainment industry, as established performers from vaudeville, theater, or film would not allow their personas to be completely overtaken by a corporation. Established people had gone by their stage names and, understandably, wanted to continue to capitalize on those names. Although this allowed sponsors to advertise in a commercially sensitive environment, this situation further discouraged the entry of established talent.

As arguments over radio's commercialization began to die down in the late 1920s and as larger corporations began investing more money into radio advertising, the nature of radio talent and formats underwent considerable alteration. An early sign of this was the move toward hiring name talent. One way that agencies provided a strong distinguishing element to their programs, and thereby potentially guaranteed large audiences, was to employ presold, top-name talent from other entertainment media. An advertising executive and writer for *Chase and Sanborn*, Carroll Carroll recalled that, "the real gut power of radio surfaced around 1931 when advertisers began to abandon such obvious broadcast nomenclature as the A&P Gypsies, Paul Oliver and Olive Palmer in the *Palmolive Hour*, the Gold Dust Twins, the Happiness Boys (later the Interwoven Pair—a sock act), the Clicquot Club Eskimos and [replace them] with the use of star talent."[13] There was also a move toward the production of fictionalized narratives, such as serials, anthology dramas, musical programs, and

Fig. 1.1   The Happiness Boys (Billy Jones and Ernie Hare). Library of American Broadcasting.

variety shows, which as Michele Hilmes points out, came out of broad-casters' need to have regular and reliable output of product.[14]

Many radio historians consider The Fleischmann's Yeast Hour with Rudy Vallee to be the first program to signal the medium's transition to variety programs and pre-sold talent and single out Vallee as radio's first real star. Premiering in 1929 and running for almost ten years, The Fleis-chmann's Yeast Hour had Vallee at its center, around which high-profile guest stars such as Eddie Cantor and Burns and Allen would appear in

Fig. 1.2 Rudy Vallee. Library of American Broadcasting.

comic sketches. Vallee had established his persona on Broadway and in
night clubs where he was a popular and attractive band singer.[15] At this
point, advertising agencies still used the sponsor's name in the title of their
programs but allowed their hosts and regular talent to retain their prior
names and images. These precedents anticipated radio's move toward the
poaching and recasting of talent from entertainment media, as many
Broadway performers tested out the medium through guest spots on these
early variety programs.

Vallee was a crooner, which was a popular form of romantic, electronically
amplified singing done by a young male band singer with a microphone.

Fig. 1.3 Bing Crosby. Library of American Broadcasting.

From 1929 through the mid-1930s, crooners such as Vallee, Bing Crosby, and Russ Columbo were the most listened to, and at times the most controversial, singers on the air. In her study of Vallee, Allison McCracken remarks that crooning "was remarkable for its homogenizing synthesis of American music, as it combined the intense romanticism of the Victorian ballad with the amorality of the urban novelty song and the emotionalism and sensuality of jazz music."[16] She also notes that within this musical and performative combination existed an amorphous sexuality that, although attractive to many women, led the press to deride singers such as Vallee and Crosby for being both playboys and oddly effeminate. The crooner's female fans were enraptured by the intensely romantic style as well as the seeming nearness of the singer's presence. McCracken argues that this

Fig. 1.4 Bing Crosby hawks GE radios. Interestingly, the ad promotes programs on two networks with references to NBC's *Kraft Music Hall* (of which Crosby was the star) and two GE sponsored programs on CBS, *The World Today* and *All-Girl Orchestra*.

had everything to do with the crooner's use of electronic technology, as the combination of an amplified microphone and radio broadcasting offered an intense type of intimacy that couldn't be found elsewhere—even on stage. Timothy D. Taylor agrees with McCracken and adds that the crooners were constructing an intimacy not with their audience in general, but with one (female) listener, adding that, "crooning as a singing style thus introduced a paradox: whereas radio was proclaimed as uniting disparate Americans into a single national culture, this singing style that had been ushered into existence by radio helped create and maintain an illusion that listener's relationships to singers and other broadcasting individuals were unmediated, personal."[17] Radio's placement in the domestic coupled with its liveness creates a sense of intimate oneness between listener and performer just as it can recreate a sense of place within a crowd, such as a packed theater. This interplay between individual and community experience is essential to the relationship between listener and broadcasting star. Rudy Vallee was the first to fully exploit it.

In 1929, NBC acquired the talents of two performers who would participate in the first national radio phenomenon. Freeman Gosden and Charles Correll appeared on NBC Blue on August 29, 1929, in their program *Amos 'n' Andy,* which was picked up by the network because of the local success of their WGN Chicago show, *Sam 'n' Henry.* Based on vaudeville traditions such as minstrel and "rube" acts, Gosden and Correll's program was preoccupied with racial dialect and cultural incompetence in its representation of African Americans through aural blackface.[18] In part because of its serial narrative, sitcom structure, and mobilization of 1920s racial discourse, the program was so wildly popular that some have gone so far as to credit the program with causing radio's rapid diffusion in the early 1930s. In the case of *Amos'n Andy,* however, it was not the name nor the personas of Gosden and Correll that captured the nation's imagination; rather, it was the personalities of the characters they played. It wasn't until the introduction of ex-vaudevillians to radio that a performer's stage name became essential to the development and marketing of programs.

The exploitation of vaudeville headliners and sound comedy film stars would be a primary factor in the standardization and differentiation that was established during radio's classical period. In the 1956 *Broadcasting in America,* Sydney W. Head claimed that two of the major accomplishments of 1930s broadcasting were the adaptation of entertainment formats (from vaudeville, theater, and film) to the radio medium and radio's development of star talent. Head writes:

> Radio discovered in the 1930s, as cinema had discovered years before, that successful syndicated programming on a national scale depends in large measure on certain intangible assets by star performers. These assets justify paying the star salaries...The control of talent was from the

first an important factor in successful network operations, for which reason both NBC and CBS operated their own talent agencies until the practice was condemned by the FCC. In the strategy of network competition the ability of a network to command a lineup of top stars remains as important as its ability to muster a lineup of top stations as affiliates.[19]

With a focus on top-name talent, networks could provide affiliates with something that they couldn't afford on their own and sponsors with someone whom audiences would associate with their product. The emphasis on such performers would contribute to the rise of the radio comedian and no greater investment in the variety format. Although the acquisition of individuals such as Cantor, Edgar Bergen, and Fred Allen would raise production costs to previously unforeseen heights, it would also help make the network broadcasting a national pastime and mark the beginning of the broadcast industry's reliance on vaudeville-inflected comedy in prime-time programming, a practice that would continue into television.

Ex-Vaudevillians in Radio Variety Programs: The Success of Eddie Cantor

Presold talent was not all that radio would borrow from Broadway and Hollywood. The format of the traditional vaudeville bill, with its olio structure and patchwork presentation of various musical, comedy, and acrobatic acts, was well suited to the characteristics of the medium as well as to the relationship that the sponsor desired with its audience/consumers. The variety format was highly pliable in that if a particular type of sketch or guest act bombed, producers could replace the segment the following week. If audiences were especially responsive to a specific performer, character, or sketch, a sponsor could make that a regular feature of the program. Moreover, vaudeville was a felicitous stage style for radio to ape in that broadcasting's dependence on local outlets resembles the very structure of a vaudeville circuit.

Up until the 1920s, networks were still producing almost half of what audiences heard on the air.[20] However, by as early as the 1930s, ad agencies had almost completely taken over the production of national sponsored programming.[21] During radio's heyday, the agencies developed programs in-house for sponsors and then purchased air time from networks or local stations. Although the agencies maintained a level of creative control, the sponsor held the purse strings and could therefore nix material, talent, or formats. This lack of network control affected the aims of programming narratives as well as the manner in which the audience was perceived. Agencies were required not only to establish an appropriate environment for commercial messages but also to create a program that could appeal to an audience large enough to satisfy the sponsors' need for a consumer base. Sponsored shows that did not rack up audience numbers were useless in this environment, and the resulting urgency in which agencies needed to track audience size augmented the use of radio rating services. Moreover, the program itself became a product, as agencies launched major merchandising campaigns complete with print ads, trade paper notices, displays, billboards, and broadsides.[22] The sponsors had access to larger and broader demographic numbers because of network domination of local stations. In order for a station to be successful in the 1930s, it needed to be affiliated with one of the four national networks —NBC-Blue, NBC-Red, CBS, and Mutual. In 1935, 14 percent of all radio stations in the United States were affiliated with one of the NBC networks, and NBC owned ten of their own stations.[23] Approximately 15 percent of stations were affiliated with CBS, and that network owned nine stations outright.[24] Mutual and other regional networks had agreements with local low-power outlets that attracted only small audiences, so sponsors with large advertising budgets put their money with either CBS or NBC, producing national programs with broad appeal.

Throughout radio's golden age (roughly 1934–1941), the most popular prime-time formats were the anthology (or prestige) drama and the variety show. Largely because of the revolving casts of shows such as *The Lux Radio Theatre, Mercury Theater of the Air*, and *The Screen Guild Theatre*, few significant original personalities arose from the anthology shows. Very few regular figures or characters occurred, limiting identification and advertising promotions.[25] (Orson Welles was a notable exception.) Consequently, singers and radio comedians hosting the variety programs were the ones who were constructed as the stars of the airwaves. The variety show maintained one pivotal personality—or host—who would link together a series of sketches, musical numbers, acts, and/or monologues. Usually the host was aided by at least one subsidiary character. For example, Charlie McCarthy and Mortimer Snerd accompanied Edgar Bergen; Rochester was always alongside Jack Benny; Jerry Colonna supported Bob Hope, and Fred Allen had his real-life wife, Portland Hoffa.

Cantor was one of the first Broadway comedians to make it big in radio. As host of the *Chase and Sanborn Hour*—which premiered on NBC on September 13, 1931, and was produced by one of the largest broadcast

Fig. 1.5 Eddie Cantor. Library of American Broadcasting.

ad agencies, J. Walter Thompson—Cantor performed many of his signature routines. But he also altered his style to suit the needs of a commercial broadcasting program. He discovered that instead of depending
solely on gags, the broadcast audience remembered and responded better
to characters (the Mad Russian and Parkyakarkas) and multiepisode
themes (such as his Cantor-for-President campaign).[26] He used his

announcer, who would often insert commercial references into the program, as his straight man. Even with all these alterations to his stage technique, Cantor, who was one of the most popular vaudeville and Ziegfeld Follies headliners of the 1920s, would occasionally neglect to take into consideration radio's wholly aural environment as he continued to perform sight gags and costume jokes on the air, which were funny to the studio audience but left the listening audience completely baffled.[27] Overall, Cantor handled the transition to radio quite well and would serve as a model for future radio comedians, yet figuring out exactly how a persona was signified in radio remained a bit perplexing. Writing about the psychology of radio in 1935, Hadley Cantil and Gordon Allport discuss the difficulty of fashioning a definitive personality for the radio performer: "The [talkie] is designed to give the personality of the actor the fullest display. The star of the talkie is seen and is heard, and in a close-up can almost be tasted. But the radio star is present only to the ear, and his personal qualities and appearance must be left in part to the imagination."[28]

Constructing or transforming a persona to fit radio was a crucial element in radio stardom. The ad agency could then utilize these personas in various ways. J. Walter Thompson executives knew from previous experience that celebrity endorsements could produce great profits. (Michael Mashon points out that Thompson had used this strategy since the success of its "Nine Out of Ten Stars Prefer Lux Soap" print campaign of the early 1920s.[29]) Therefore, J. Walter Thompson was more likely to develop shows that contained highly visible performers at their center.[30] That way the agency could link the star's persona to the sponsor's product within the program text as well as outside of it. By the 1930s, it was fairly common practice to incorporate commercial messages into the program's narrative. However, Mashon's research on J. Walter Thompson's radio department shows how important and yet still financially risky it was for one of the most active ad agencies in broadcasting to develop a show around one or two performers:

> The Thompson programs depended on proven star power to generate audiences, which was at once a conservative and radical approach; conservative in the sense that the agency was relying on pre-sold commodities . . . and risky in that these stars demanded enormous salaries at a time when most agencies were moving into radio with some trepidation and modest financial resources.[31]

The high cost of talent became a central issue for sponsors at the very time that Eddie Cantor began his run on radio (and would rise again during the early 1950s for television sponsors). Because advertising had taken over radio production, in 1932 the American Society of Composers, Authors, and Publishers (ASCAP) and the talent agencies decided to stop allowing their members and clients to perform without pay. During the period of early growth and dissemination of the medium in the 1920s, they had considered free performances simply good publicity for actors, musicians, comics, and announcers. But as performers and their representatives started to see commercial sponsors profiting off their live and recorded performances, they demanded fair remuneration.[32] Indeed, as Cantor's popularity increased, so did his salary. A 1941 article revealed the extreme disparity of salaries in the broadcasting industry by comparing Cantor's salary to others. It claimed that a page boy earned $18 a week, an executive made $1,000 a week and Cantor pulled in a whopping $10,000 for eighteen minutes on the air.[33] Even with the high cost, Cantor was a sound investment for Standard Brands, since at the time of entry into radio he had already proven his star power in two other media. Cantor was one of the most successful comedians on Broadway in the 1920s. *Whoopee*, a Ziegfeld production with Cantor in the lead role, was the top grossing musical during the 1928–1929 season.[34] Cantor's films *Kid Boots* (Famous Players, 1926) and *Whoopee* (United Artists, 1930), although not as widely successful as his Ziegfeld performances, drew in significant box office numbers. The persona that Standard Brands was buying through its investment in Cantor's fame, however, was one that would primarily appeal to a New York audience. Henry Jenkins notes that after years of performing for Ziegfeld, Cantor had become closely identified with the sensibility of New York City:

> Cantor enjoyed the freedom of a more "sophisticated" following; his comedy was sprinkled with cynical remarks about marriage and with eye-rolling double entendres that even hardened local critics sometimes found "too blue" for their tastes. Moreover, Cantor relied heavily on his Jewishness as an appeal to the city's ethnically diverse theatrical audience.[35]

The context for Cantor's urban, Jewish sensibility grew out of what Albert McLean calls vaudeville's "new humor"—a style that came into practice at the turn of the century along with what Douglas Gilbert calls "Jewish character comedy." The "aggressive" and "excited" wordplay of the new humor resulted in part from interaction with vaudeville's new immigrant audience. McLean writes that "by 1900 both characters and situations had become stereotyped and standardized to the point where the only novelty lay in the fluid, living language of the cities—in dialect, in boners, in slang, and other surprises of sound and syntax."[36] Often the source of the style's characterizations and verbal play was the stereotyped image of the Jewish immigrant. Although the verbal play of this type of humor would suit the purely aural nature of the broadcasting medium, much of the content would have to change. As Susan Smulyan notes, "Performers coming to radio from vaudeville . . . lacked experience with the restrictions imposed on home entertainment—experience earlier radio performers had already gained in the recording industry."[37] Consequently, the blue humor and ethnic references that characterized Cantor's Broadway style—although appealing to much of the urban radio audience —would need to be sanitized and homogenized by producers and network censors for *Chase and Sanborn*'s national NBC audience. Nevertheless, radio comedy would retain some elements of its vaudeville origins, including, in Cantor's case, ethnic humor.[38] These experiences with radio censorship would also eventually have significant influence in decisions television producers would make in their mediation of vaudeo content and fashioning of personality.

Although he began his career in radio as a gag comedian, by 1934 Cantor was openly criticizing the formula that he had perfected on Broadway and in his first few years on radio. At one point, he told an interviewer that he'd "like to get away from gags altogether . . . the public is sick of them. We rewrite old jokes, dress 'em up, and call 'em new gags. But you can't fool the public."[39] Critics were tiring of the slapstick and gag routines as well. In 1935, A. M. Sullivan remarked, "For the past seven years radio broadcasting has been operated on the vaudeville formula, and even the ex-booking agents of 10–20–30 cent circuits recognize that the pattern is getting threadbare."[40] The concern over the content that vaudevillians brought to radio from the stage continued to plague the format throughout the 1930s and was often tied to the criticisms of gag comedy.

Although criticism of the gag and variety format existed, the popularity of *Chase and Sanborn* led to a proliferation of shows of a similar ilk. In 1932, shows featuring Fred Allen, George Jessel, Jack Benny, Ed Wynn, and Burns and Allen, premiered on prime-time network radio. Broadcasters recognized the saliency of the hosts' personas and found that their high visibility sustained a show's top ratings. In exchange, radio provided ex-vaudevillians with the opportunity to achieve a fame that reached far

Fig. 1.6 A display of Eddie Cantor merchandise and publicity material. Library of American Broadcasting.

beyond that which they had known before. Yet the sponsors and networks would often end up in battles over content and behavior with their stars. In particular, network executives were on the lookout (particularly during World War II) for transgressive political behaviors and opinions. Cantor, who was very outspoken about his religious and moral beliefs during his off-air time, was often chided for his actions, even as his charity work was applauded by the press. One of Cantor's scrapbooks revealed that he had donated his time to the Red Cross, the Army and Navy Relief Fund, Hadassah, and the Children Refugee funds as well as various old-age homes, Catholic charities, and orphanages.[41] He also was an active Zionist.[42] A Young and Rubicam memo stated the reasons they thought that Cantor's popularity had dropped off in those years. The memo is revealing in that it outlines the agency's belief that comedy is incongruous with intellectual or social causes:

> (1.) in [his last radio show] if he didn't flop, he didn't ring the bell. (2.) His last picture was a flop. (3.) His appearance at the Capitol Theatre was a flop. (4.) His non-comedy activities have tended to present him

as a serious minded person, making it difficult to appreciate him as a person to be laughed at. (5.) His attack on Father Coughlin was ill-advised. (6.) The manhandling of two studio guests left an ugly impression. (7.) His charge that radio editors lack honesty of purpose received unfavorable attention.[43]

Despite his declining popularity in the late 1930s, Cantor set the tone for radio variety programming. His early success was convincing evidence of the compelling nature of the radio vaudeville persona. Once some of the brash or outlandish aspects of vaudeville performance were quieted, the form would fairly consistently please audiences, critics, and sponsors. Still, the very debates and issues surrounding the emergence of radio comics such as Cantor would be recapitulated years later in television, showing the difficulty, and perhaps futility, of completely sanitizing this generic form in broadcasting.

Hollywood, Radio, Stardom

The relationship between broadcasting and Hollywood has been a complicated one from the very beginning, as it has been riddled with both ambivalence and hopeful expectation. There was much potential for cross-promotion but also for much competition. However, much of what would occur during the 1930s among the entertainment industries would not only establish long-standing financial relationships but also would lay the groundwork for the way in which broadcast stardom (both radio and television) would be culturally and industrially imagined in relation to film stardom.

Hollywood has been a talent poacher in much the same way that radio and television have. In fact, after finding many of its stars in vaudeville, nightclubs, and legitimate theater, studios tried to make film stars out of popular radio performers. In his comprehensive history of 1930s Hollywood, Tino Balio argues that there were two ways radio stars were used in the movies during that time: (1) they were cast as the central character and the film was written as a star vehicle for them; or (2) a backstage musical narrative was used as a way to showcase a number of radio performers doing their regular routine.[44] The first strategy only really worked for one radio star—Bing Crosby. The second strategy was more successful generally and resulted in a number of fairly popular films, most notably those of Paramount's Big Broadcast series (*The Big Broadcast*, *The Big Broadcast of 1936*, *The Big Broadcast of 1937*, *The Big Broadcast of 1938*) which starred such luminaries as Crosby, Hope, Benny, Burns and Allen, Kate Smith, and Cab Calloway as themselves in films set in the radio industry.[45] Overall, however, radio stars did not do that well on film and producers had a hard time explaining exactly why. It seemed as though film stars had an easier time adapting their performance conventions to the narrative world of radio (particularly when it came to radio adaptations of Hollywood films) than the other way around. As a writer for *Variety* stated in a review of one radio team's recent picture, "a whole list of radio folks who went to Hollywood, made one picture, and apart from a piece of change, did themselves little good."[46]

The same year that vaudevillians took over the airwaves, 1932, marked a turning point in the attitudes of film studios with regard to the potentialities of radio promotion. Instead of seeing radio as simply a competitor, studio publicity departments began to regard radio a lucrative and creative venue for their marketing campaigns. Some studios such as RKO-Radio (which was owned by RCA) and MGM even began sponsoring their own radio programs during that year. Most studios, including RKO and MGM, focused their efforts on promoting new releases through backstage gossip shows or movie adaptations.[47] Although stars were not charging studios for their appearances on such shows, some in the industry noted that there might be other long-term costs to exploiting film stardom in the domestic medium of broadcasting. Exhibitors had long been upset by the studios' lenient attitude toward these appearances and, in fact, enacted a boycott of radio in 1934. Although, as Hilmes proves, the length and strength of the Radio Ban of 1934 has been generally exaggerated by historians, it did force studios to at least give voice to the negative, yet still quiescent, effects of this type of exploitation of stars. In addition, it allowed studios to exert more control over the appearances of contract talent.[48] A complicated relationship developed among studios, exhibitors, and broadcasting during the 1930s as studios increased their direct and indirect investments into radio and a significant number of exhibitors actually espoused the advantages of radio as a publicity tool for Hollywood. Another interesting by-product of this struggle over the appearances of contract talent was the revelation of the industry's working assumptions about the nature of film stardom and how broadcasting might impinge on it.

The ex-vaudevillians appearing on radio during these years were willing to be employed by advertising agencies and to engage in the process of on-air selling. At least at first, legitimate stage and Hollywood film stars were far less malleable in this regard. Screen stars either did not wish to be "tainted" by the brash commercialism of sponsored programming or were already beholden to the wishes of a studio. Sometimes they would appear on radio variety shows as guest stars or participate in dramatic programs or movie adaptations to promote studio product. Studios and exhibitors were greatly invested in their stars' images and were therefore quite protective of them. The exhibitors feared that film stars' repeated exposure on radio would keep filmgoers at home as well as lead to a depletion of their box-office appeal. Specifically, they were angry that stars were giving "free" performances during prime movie-going hours. Studios, however, were less worried about radio's deleterious impact on the elusive nature of film stardom and more interested in regulating star radio appearances to their own advantage.[49]

The ontology of the movie star as delineated by the industrial discourse of the 1934 radio ban hinges on the idea that a star's presence or image is a limited commodity. An image can be "spent" through an appearance in a film, publicity junket, or radio broadcast. Although an outside observer might think that any amount of publicity would be good for a star's box office appeal, the industry believed otherwise. It would seem that, in order to get spectators in the theater to see a specific star, that star should have had restricted circulation in media environments. Interestingly, articles in fan magazines may not have been a part of this equation since they were constructed in a manner simultaneously to reveal and conceal information about the star in order to pique a fan's interest.[50] As Cathy Klaprat has discussed in her work on the film star as market strategy, each industrial tier—production, distribution, exhibition—had individual but interlocking uses for the star image.[51] Producers used stars in order to differentiate product and to create a house style for the studio, whereas distributors based rental prices on the known drawing power of a specific star, and exhibitors used stars to attract audiences to their theaters.[51] Overexploitation of the star image could cause a decrease in his or her rarity value and, therefore, potentially threaten the system of standardization and differentiation engaged across the three tiers.

Exhibitors were especially reluctant about the appearance of film stars on radio. The fact that a star would "expend" only the aural aspects of his/her image through the medium makes the exhibitors' anxiety about radio in particular a bit curious. Yet because exhibitors considered broadcasting to be a direct competitor with film, their wariness of the new medium in general may explain a large part of their concern. Aligning radio appearances with the depletion of star rarity value was a clever rhetorical strategy for the exhibitors in their struggle with studios, as depletion could potentially have had reverberations throughout the entire industry. So, by using it as the premise for the ban, the interpenetration between broadcasting and Hollywood could have been successfully questioned.

But what does this struggle tell us about broadcasting's perceived impact on the circulation of star images? To begin with, it speaks to the way in which studio-era Hollywood distanced itself rhetorically from the machinations of mass production and commercialism via its comparisons with the broadcasting industry. Even though Hollywood obviously maintained the economic form and structure of an industry engaged in mass production, it was successful in creating an image of artistic endeavor. This is not to say that one could consider Hollywood output as high culture rather than popular, or mass entertainment. Rather, it is to point out how Hollywood was able to place itself above what was popularly considered the "crass" commercialism of broadcasting. The production of narrative by advertising agencies, the ubiquitous presence of the sponsor's product on the air, and the massive, immediate output of the medium left the broadcasting industry vulnerable to such accusations. Even though broadcasters were successful in their campaign for commercialism in the 1920s (primarily through allusions to national unity through networks), they were never able to completely squelch the often accusatory rhetoric (on its purported social, cultural, and psychological effects) circulating around the uneasy alliance between advertising and entertainment. In the Golden Age, this critique took the form of campaigns for educational programming and "quality" and cultural uplift in, at least, network sustaining programming. According to the rhetoric deployed in the Radio Ban, the Hollywood star also was defenseless against the effects of radio's commercialism. Stars could be tainted, reduced, or possibly destroyed by their engagement with the medium.[52]

The discourse surrounding commercialism and film stars versus radio stars reasserted a hierarchical relationship between these forms of stardom. It asserted the preeminence of film in the production of "high-quality" stardom and placed broadcast stardom in a secondary position. This not only affected the broadcast industry's ability to attract Hollywood talent to the medium but also set up the assumption (still at work in star theory today) that television stars were not as culturally worthy or potent as their filmic counterparts. Yet the very feature that defines their so-called inferior status—their overt commercialism—was the thing that made their images so culturally prolific and influential.

An Audio "Star-System": Star Branding, Continuity, and Cross-promotion

J. Fred MacDonald points out that, "[w]hile the great movie studios at this time were using contracts and business agreements to create the so-called 'star system' of motion-picture celebrities, in radio it was the advertising agencies that were developing an audio star system."[53] Consequently, radio's star system (and, later, television's) had different contours from that of screen or stage. The most glaring difference was the decentralized nature of radio's system and the way in which it had to manage the various commercial and program texts in which a potential star would appear. Most important, the advertisers and networks had to create continuity in their stars' personas in order to ensure uniformity between the star and the product brand. The lessons that agencies and networks learned here would have a major impact in how they envisioned and eventually packaged postwar television stars.

Unlike the studio-era film star who was required to represent a particular studio and its products, the radio star was committed to represent the sponsor, the network, and, to a lesser extent, the advertising agency who employed him or her. A high-profile radio performer would sign a contract with the network and the advertising agency, releasing the use of his or her image for advertising purposes. Thereafter, the performer would often engage in cross-promotion, appearing in print ads, radio spots, and billboards peddling consumer products, the program, the network the program aired on, and sometimes even the network's radio sets as was the case with NBC. If another company wanted to hire a radio star for an ad campaign, contracts required the hiring firm to carry the name of the show's sponsor, network, and its time slot. Additionally, consumers might have found their favorite radio star in a film at their local movie theater. This venue provided a broadcast performer broader exposure and a certain level of prestige that would benefit the performer, network, and sponsor. Although it was assumed that too much radio exposure could cost a film star some of his or her box-office draw, it was generally believed that success on film would *add* to a radio comedian's profile with audiences.

What is especially intriguing in relation to the project of selling the star who began in radio is that the initial textual image is constructed in a purely aural medium. Thus, the secondary circulation of image (in print ads, films, etc.) is essential to the potency of the initial one. The visual construction needs to coincide with the star's on-air image yet also provide some essential information to fill out the audience's imagination of who

Gracie bulbsnatches while George Burns!

Nobody knows better than you, Gracie Allen, that George Burns is a patient man. But *bulbsnatching!* ... what red-blooded head-of-the-house can take *that* with a smile? Robbing one light socket to fill another just leads to strained eyes, barked shins, endless annoyance.

No wonder George is hot under the collar! How to cool him off? Easy—get some of those bright, dependable G-E lamp bulbs. They cost so little, any budget can afford them. And they're backed by G-E lamp research, which is constantly at work to make G-E lamps *stay brighter longer!* That means more and more light for your money.

So, take a tip from Gracie, folks. Keep your tempers in check by keeping G-E lamps on hand. See your G-E dealer today!

G·E LAMPS
Stay Brighter Longer!

25, 40, 60 WATT
100 WATT 15¢
150 WATT 20¢
100-200-300 55¢
plus tax

GENERAL ⊛ ELECTRIC

Be sure to hear **BURNS & ALLEN** on the **MAXWELL HOUSE COFFEE SHOW** every Thursday over NBC at 8:30 P.M., E.S.T.

Fig. 1.7   George Burns and Gracie Allen promote their radio program as they sell GE lightbulbs.

this particular radio performer/character is. One would have to be exceedingly conscious of the maintenance of image continuity especially if the star was involved with cross promotion. In order to achieve continuity both within the text and outside of it, an agency would first attempt to maintain continuity in the form and content of the program. Felix describes the role that textual continuity played in the hiring of talent:

Fig. 1.8   Charlie McCarthy and Edgar Bergen for GE.

Continuity or sustaining power is in a sense the direct opposite of attention-compelling power. Having won attention for a feature, the next thing which must be accomplished is to make all those who hear it regular listeners. This is accomplished by the pursuance of a definite program policy, employing the same sustaining artists. Guest artists may augment the permanent group, but it is of great importance that the principal artists be retained regularly.[54]

So, although continuity within the program was important to keeping a regular audience of listeners, in order for the ad agency to conjoin the "sustaining artist" with the sponsor, continuity within a star's textual and extratextual performances was also a priority. Beyond continuity with the program, the star's persona had to be consistent across varied media. For this to be achieved, the character that the comedian played on his or her radio program had to at least partially resemble his or her "real-life" personality. For example, Benny maintained recognizably consistent characteristics on the air, in the publicity on his personal life, in his films, and in print and broadcast ads. On a very basic level, in the case of most Golden Age radio stars, the program's title and main character all bore the stage name of the performer. This was because sponsors recognized that the name of a vaudeville headliner was the primary hook in the effort to attract audiences and to differentiate the program from the old-style orchestral shows. Because the star's name (and persona by implication) was the most recognizable and therefore exploitable name associated with the program, it was the program's brand name. Moreover, for example, Benny had to remain "Jack" across media in order to preserve a continuity of image. The sponsor's goal in this process was to make their product's name either synonymous, or at least easily associated, with that of the performer's name.[55] As a result, the nature of a performer's character and performance style had to be constructed in a manner that coincided with the product's image. Felix vigorously warned sponsors that:

> If dependence is placed upon a personality, such as impresario or announcer, to establish continuity and sponsor consciousness, extraordinary care must be used in selecting the individual for this role. Any voice failing, such as foreign accent, whining or nasal quality, affectedness or effeminateness, must be positively avoided.[56]

A factor complicating the project of continuity was the radio star's perceived authenticity in regard to his or her off- and on-air lives. Although cinema-star studies have established the intuitive and analytical abilities of fans to understand the deliberate fashioning of star images by the industry, it is also thought that much of a fan's pleasure in engaging with extratextual materials comes out of his or her search for the authentic persona behind the construction.[57] By the time the variety and sitcom formats had been set in radio's Golden Age, the formula for a successful negotiation of a radio performer's "real" life versus on-air life was also established. Many of the radio comedians during this period not only used their stage name as their radio character's name but also employed certain aspects of their personal lives for use in their sketches and monologues. Benny, Allen, and Burns all played off their relationships with their real-life wives in their programs. Mary Livingstone was initially only the name of the character that Benny's wife, Sadie Marks Benny, played on the program, but she adopted it as her legal and professional name once the show became a national phenomenon. In addition, the rest of Benny's supporting cast (except for Eddie Anderson, who played Rochester) performed under their own names. Benny's negotiation of direct showmanship and fictional characterization is an important point in the development of commercial and narrative continuity. Benny played basically himself in a backstage and at-home narrative based on an approximation of life in show business. Because of this backstage structure (akin to that of 1930s musicals), Benny could engage directly with his audience, introduce musical acts, sketches, and monologues, and yet remain within the framework of a situation comedy. This enabled his star image to be at the center of the program while still reaping the narrative benefits of the sitcom and variety formats. Continuity for sponsor identification was preserved through the cross-media retention of Benny's persona.

Star branding also promoted continuity between program and commercial texts. Much of the trade press during the late 1920s through the 1930s was focused on the best way to manage the medium's "flow." Scheduling and programming became vital issues during this period as did the negotiation of commercial messages within a program's narrative trajectory. In an address to the National Association of Broadcasters in 1929, Batten, Barton, Durstine, & Osborne (BBDO) partner Roy Durstine lambasted broadcasting programmers for ignoring the connection between the commercial and editorial spots of a program. Mashon writes that "what Durstine had recognized quite early in the game was that successful broadcast advertising depended on flow, a continuity not only between the show and its commercials, but between programs as well."[58]

Although the host was most often placed as the locus of the incorporation of the sponsor's name or product, sometimes, as was the case with Don Wilson in the Jack Benny program, an announcer would act as the primary integrator of the sponsor's message. In this role, performers such as Wilson were required to insert nonchalantly into the program lines such as "I shopped around until I found half a dozen neckties, each one corresponding in color to a different flavor of Jell-O. . . . You know, Strawberry, Raspberry, Cherry, Orange, Lemon and Lime."[59] Yet even while Wilson was pushing the sponsor's name whenever possible, Benny's image became synonymous with a number of the products. In particular, the names Jack Benny and Jell-O (which sponsored the show from 1934 to 1942) became entwined in the minds of the radio audience. Not only would Benny make references to his sponsor's product—as in his frequent opening line, "Jell-O again. This is Jack Benny"—but Benny was required by his contract with the sponsor to allow his image to be used in ads outside the confines of the program's time slot. This also was true of the majority of the successful radio comics. (Bob Hope, for example, would start his show by quipping, "This is Bob 'Pepsodent' Hope.") In successfully making this connection, an advertising agency could unite the merchandising of the program with the project of selling product; retain the connections among network, star, and product throughout an entire campaign; and naturalize the incorporation of references to the sponsor's product with the program's narrative structure. The blending of commercial and program narratives provided a justification for the sponsor's production of the entire program and located the star as the center of both narrative trajectories. In selling this advertising strategy to agencies, NBC, according to Hilmes, "argued that this 'interweaving' technique gave the advertising message added force and credibility and avoided the potential 'offense' in more direct selling messages: listeners would scarcely be aware they were being sold."[60]

Maintaining continuity of the star's image mattered not only to the performer and the product sponsor but also to the network. The manner in which networks could utilize star performers is similar to the way studios used film stars to promote a certain "house style." For example, NBC clearly believed that Benny's image was integral to their creation of a specific network image. By the early 1940s, NBC had given Benny the time slot of Sunday nights at seven, no matter who was sponsoring his program. Although the sponsor owned the time, that time slot on NBC had become so associated with Benny that the network saw it in its own best interests to make Benny this unprecedented promise. NBC's reliance on stars was clear by this time as they maintained the largest stable of names in the business, including Bergen, Godsen and Correll, Red Skelton, and Burns and Allen. NBC controlled the circulation of the star's image as best they could in order to stifle competition; for example, NBC attempted to restrict guest performances of its stars on the other networks. Moreover, the voracity of CBS's talent raid in 1948 also speaks to the perceived importance of the links between particular stars and their networks.

Imagine Bob Hope... on **TELEVISION**

brought to you by **N B C**

Think how, on NBC Television, the amusing antics of America's greatest comedians . . . the zany adventures of Bob Hope, for example . . . could take place before your eyes in hilarious *visual action*.

Just picture how television programs from the studios of the National Broadcasting Company . . . where the nation's most popular sound radio programs now originate . . . are going to up the excitement of home entertainment.

At the present time, NBC has extensive television plans under way. With the co-operation of business and government these plans, after the war, will bring about vast NBC Television networks . . . networks

gradually sprouting from Eastern, Mid-Western and Western centers and finally grouping together . . . forming coast-to-coast links to provide television for the whole nation's post-war enjoyment.

Popular-priced television receivers will bring to your home sight and sound programs up to the highest standards of NBC . . . television programs of the finest shows in this fascinating and improved field of entertainment.

*   *   *

You can depend on NBC to lead in new branches of broadcasting by the same wide margin that now makes it *"The Network Most People Listen to Most."*

National Broadcasting Company

America's No. 1 Network

A Service of Radio Corporation of America

Fig. 1.9 NBC asks its audience to "Imagine Bob Hope on Television" in 1944.

The Saliency of Broadcast Stardom

CBS's move to lure NBC's most important performers over to CBS was preceded by a decade of shifting relations among performers, talent agents, sponsors, and the networks. NBC and CBS, believing they could garner some control and financial benefit from representing performers (and eventually writers and directors), had established their own artist bureaus in the 1930s. However, under pressure from the government by 1941, both networks had sold these divisions. The Federal Communications Commission's *Report on Chain Broadcasting (1938–41)* that focused on NBC found that:

> As agent for artists, NBC is under a fiduciary duty to procure the best terms possible for the artists. NBC's dual role necessarily prevents arm's length bargaining and constitutes a serious conflict of interest. Moreover, this dual capacity gives NBC an unfair advantage over independent artists' representatives who do not themselves control employment opportunities or have direct access to the radio audience.[61]

Soon after the report came out, the CBS Artist Bureau was sold to MCA whereas the NBC Artist Bureau became a new agency, National Concerts and Artists. MCA went on to become the most powerful talent agency in motion pictures and broadcasting, representing performers, directors, commentators, and writers. Because of the government's intervention, creative talent would have representation that would look out for their interests instead of those of the network and sponsors. A direct consequence of this would be a rise in salaries and fees as well as a slow swelling of star power within the hierarchical structure of broadcasting.

Why did this occur? Part of the answer lies in the significant cultural and social influence that top name talent were accruing. One of the most striking examples of the radio stars' power was the way in which they were used for the war effort in the early 1940s. During World War II, the Office of War Information (OWI) enlisted radio performers to aid in the war effort. Hope, Bing Crosby, and Benny were extremely active in the entertainment of troops stationed in Africa, Italy, Iran, and the South Pacific. Others such as Burns and Allen, who performed on a War Bond Campaign show, a Navy Enlistment broadcast, and a China Relief Program, appeared as guests on special war-related radio programs.[62] The most common practice, though, was the inclusion of war-related material in the text of a comedy program or of spot announcements at the show's conclusion. During the war years, writers added an additional layer to the technique of interweaving commercial and program narratives as they introduced government messages on patriotism, rationing, and the draft into the texts. And, once again, stars eased the transition from entertainment to directive. Crafting the war messages at the OWI's Domestic Radio Bureau, former advertising executive Donald Stauffer headed up the group. Stauffer even claimed that stars helped to dress up their war messages "in six delicious flavors."[63] Radio stars' participation in the war effort had a return effect: it endowed them with increased potency as national symbols, as the OWI constructed them as essential tools in the propagation of nationalistic ideals and wartime conservation. In some cases, they were seen as so valuable to the war effort that they were recognized by the government as comedic heroes.[64]

By the mid-1940s, many top radio comedians were grossing as much as $25,000 a week but were still not satisfied with their take-home income or their treatment by their network (in particular, NBC). Benny and Cantor, whose shows had spawned stars and spin-offs that profited the network, were aware of their popularity with audiences and began to use that power as a bargaining point. NBC, however, made the initial mistake of underestimating their stars' intrinsic role in the construction of its schedule and network identity. In 1946 contract renewal negotiations, Crosby requested that he tape-record his Kraft radio series so that he could edit out mistakes and broadcast his best singing performances. NBC refused, fearing that recorded programs would lead to the demise of live network broadcasting. The network also claimed, because it was required to use only RCA equipment and RCA had not developed a viable recording machine as of yet, that it also was technically unable to meet such a demand. ABC, however, was willing and able to use an APEX recording system and lured Crosby, and eventually other performers, over to the network with the promise of taped programming.

During the 1947–1948 season, other NBC stars demanded that their shows be recorded. Many of them felt that recording their programs would afford them more control not only over their performance but also over their lives and schedules. They were also aware that recordings could permit rebroadcasts and, consequently, additional revenue. In a December 1948 letter to NBC executive Sidney Strotz, Ozzie Nelson wrote, "I hate to keep after you like this but don't you think NBC is foolish to hasten the demise of radio by putting unnecessary restrictions on performance? . . . With the other major networks allowing tape recording, NBC is putting an unfair burden on people like us who are trying to do a good job for you as well as ourselves."[65]

CBS President William Paley recognized this moment as an opportunity to woo radio names to CBS and openly advertised the network's willingness to prerecord comedy programs. By January 1948, NBC had no choice but to capitulate. Yet, as a *Variety* article on the subject pointed

Fig. 1.10 ABC proclaims "Funny Business is our Business" in this 1947 ad. Although CBS and NBC were clearly beating the network in this regard, they had recently acquired Bing Crosby due to CBS's refusal to let him tape record his program.

out, other stars were actually against the move to recorded programs. The article reported that Benny "says he'll stay live regardless; that he's being paid for a topical show and doesn't think it fair (either to the sponsor or to the listener) to record a show three or four weeks in advance. As far as Benny is concerned, it's bad showmanship and he maintains that

Fig. 1.11 William Paley (center) with Freeman Gosden and Charles Correll. Library of American Broadcasting.

inevitably it must hurt the stars."[66] If Benny wasn't interested in moving to a network that would record his shows, salary advantages were important to him. During the 1946–1947 season, after paying his cast and team of writers, Benny was averaging $2,000 per episode for his Sunday night program.[67] Yet he was not happy with his salary and had, in August 1947, lost out on a major contract dispute with his sponsor American Tobacco. In the past, Benny had received a yearly $250,000 "exploitation-promotion-publicity fund" managed by his hand-picked publicist, Steve Hannagan, during the star's years with Jell-O and in the first years with American Tobacco–Lucky Strike.[68] Because of a turnover in management, the network refused to continue this promotion package. NBC continued to publicize his program, as did Foote, Cone & Belding, who produced the American Tobacco program. Moreover, Benny had already set up a production company of his own—Amusement Enterprises, which handled radio packages (including Jack Paar), legitimate theater, and films. CBS had never developed a strong talent lineup in radio and had eyed NBC's roster of names with envy for many years. By 1947, the network was aware that stars such as Benny, as well as Gosden, Correll, and

Fig. 1.12   CBS announces that "Jack Benny is Now on CBS" in this 1947 ad.

Bergen, were dissatisfied with their take-home salaries. Because of new income tax rates established after the war, an individual earning over $70,000 a year was required to give the U.S. Treasury department ninety-one cents of every dollar he/she earned. [69] This left most successful radio stars with rather dismal paychecks. With the help of MCA President Lew Wasserman, CBS developed a strategy to overcome the tax burden. They realized that shows such as *The Jack Benny Program* could be considered properties which, when sold, would benefit from the low capital-gains rate. The stars of the programs would be considered the properties' major assets. Although this had been common practice for business corporations and a practice in the film industry for several years, it had never been tried in any broadcasting arena. If it worked here, CBS would retain a level of control over content and scheduling, and the program's star would more than quadruple his or her net income, as CBS, on purchasing a program, also would provide stars with regular salaries for their performances.

CBS first purchased *Amos 'n' Andy* in September of that year and by December Benny had sold his program (for $2,260,000) to the network as well.[70] This was a major coup for the network as Benny (along with Hope) was considered the most popular comedian of radio's era. When CBS first approached Benny with the capital gains idea, he had gone to NBC executives to see if they would buy Amusement Enterprises from him, but

Fig. 1.13   In 1948, NBC shows that Paley hasn't taken away all its talent with this five page ad.

NBC refused, not foreseeing the significance of what would be known as the "CBS Talent Raid of 1948." As Benny and MCA began to negotiate seriously with CBS, NBC came up with a counteroffer, but it was too late; Benny was insulted by the way that NBC had treated him in their early discussions of the matter. RCA Chairman David Sarnoff had little interest in negotiating with talent, in part, because he worried that these types of capital gains transactions would not go over well with the IRS, but he was also dubious about a broadcast star system, fearing it would give talent too much power.[71] Meanwhile, William Paley personally courted Benny.

Fig. 1.13 **Continued**.

He was acutely aware of the fact that a strong star lineup could lead his network safely into the television age. *Variety* noted this in a November 1948 editorial on the capital gains deal:

> [T]hese selfsame execs are in the process of ladling out unprecedented coin for personalities; a chunk of coin, in the case of Jack Benny, perhaps undreamed of in any sphere of show business. Obviously, the primary consideration isn't in protecting a hold on Benny as a strictly radio property—but in looking to his inevitable segue into television in the era

NBC Advertisement cont.

DRAMA that has unlimited scope — from the Broadway hits and great stars and plays of The Theatre Guild on the Air and the thrilling stories drawn from the nation's life on Cavalcade of America to the tense mysteries of Mr. District Attorney, Big Town, Sam Spade, Big Story, Richard Diamond and Dragnet . . . Romance and fantasy alternate with melodrama on Curtain Time . . . Movie hits are recreated delightfully on Screen Guild, Screen Directors Playhouse and Hollywood Star Theatre . . . Powerful drama lives again, drawn from the world's classics, on NBC University Theatre . . . The saga of an American home is the story of One Man's Family — and then there are the true-life dramas of This Is Your Life.

A WORLD OF MUSIC from the bright rhythms of Your Hit Parade to the majestic beauty of the NBC Symphony under the distinguished leadership of Maestro Arturo Toscanini . . . The whole range of musical taste from the rural hoedowns and folk songs of Grand Ole Opry to the varied favorites on the American Album of Familiar Music and the Kay Armen Show, the musical delights of Broadway on the Railroad Hour, the stirring orchestrations of Band of America. Great concert artists appear on The Telephone Hour, The Voice of Firestone and Harvest of Stars. And for lighter moods the great singers of popular ballads are Frank Sinatra, Perry Como, Morton Downey, Mindy Carson and Dorothy Kirsten.

Fig. 1.13 **Continued**.

of coast-to-coast TV programming, when radio, it's recognized, will be but a secondary offshoot of video.[72]

By the end of 1948, CBS had also procured Bergen, Skelton, and Burns and Allen.[73] The loss of these programs and personnel was highly detrimental to NBC, yet Sarnoff remained steadfast in his position against offering his stars capital gains deals. An NBC executive remarked to Sarnoff's biographer that he believed that Sarnoff's resistance to making such offers was because of personal feelings of betrayal, "[Sarnoff]

NEWS from the famed NBC staff of ace reporters and commentators—men like H. V. Kaltenborn, Robert Trout, Richard Harkness, Ray Henle, Morgan Beatty, John Cameron Swayze, and W. W. Chaplin—ready to report any news break, from tiny American hamlet or from international capital; heard regularly on such broadcasts as 5-Star Extra and News of the World.

QUIZ SHOWS varying from the fabulous new entertainment jackpot Hollywood Calling to the exciting Break the Bank, and wise Dr. I.Q. to the Quiz Kids.

SPORTS covered by top notch writers, called by men like Bill Stern and Clem McCarthy. For instance, Football is in the air, and NBC brings you exciting, complete games of major colleges each week.

PUBLIC AFFAIRS bringing you stimulating information on matters of personal lives, national affairs and world problems. Such programs as You and the U.N., University of Chicago Roundtable, Living 1949 and America United.

RELIGION served by programs dedicated to our major faiths: The National Radio Pulpit, The Eternal Light, The Catholic Hour.

And, all through the day—stories, music, variety to fit your mood. *Morning to evening, tune to NBC.*

Here are more NBC Stars you know:
11. *Art Linkletter*
12. *Dennis Day*
13. *Bob Young*
14. *Bert Parks*
15. *and 16. Dean Martin and Jerry Lewis*
17. *Ralph Edwards*
18. *Frank Sinatra*

Fig. 1.13 **Continued**.

resented the attitude of the performers whom, after all, we had helped build up into stars. He felt, I think, that they were being 'disloyal.' . . . Some of us thought of talent as a marketable commodity. . . . In the final analysis we have had to accept the star system and live with it anyhow."[74]

During the late 1940s, the significance of the broadcasting star was at once buoyed and challenged by the coming of television. Most top comedians announced their industrial and cultural importance in order to secure their place (and their authority) in the unknown terrain of a visual broadcast medium. Yet with cultural and industrial recognition of their

Fig. 1.13 **Continued**.

centrality to the national network system came added responsibilities and burdens. Their ideological reach and influence would make them ready targets for the public red-baiting of the Cold War era. For example, by 1950, a significant number of the individuals who made up broadcasting's top-tier talent would be under investigation by the anticommunist organization Counter-Attack. Moreover, the transition to television would be a stressful one for most radio comics. For some, it would mean the end of their careers as the requirements of television production were taxing and deviated too much from the style of radio performance. For others, television would allow them to remain on the air for the least another decade.

CHAPTER 2
"A MARRIAGE OF SPECTACLE AND INTIMACY"

Modeling the Ideal Television Performer

With bloodshot eyes, I watch this ogre night after night, bored but nevertheless fascinated by its potentialities. How long can I survive on radio against this new monster?

—**Groucho Marx**[1]

During the late 1940s and early 1950s, as television was gradually becoming the dominant form of domestic entertainment in the United States, broadcast networks, sponsors, advertising agencies, talent unions, talent agencies, and the audience actively renegotiated the meaning and functions of broadcast stardom. Working from the premise that TV can be as good as the talent it presents, the trade press debated what qualities and experience a television star should possess.[2] Although a brief flurry of discussion on the potential advantages of the legitimate stage actor's crossover to the nascent medium occurred in 1947–1948, eventually commentators assumed that the stage comic (trained in vaudeville, burlesque, and night clubs) would be best suited to television work. This was primarily because of the trained comedian's ability to maintain the intensive schedule of television, his or her penchant for improvisation in live work, and, of course, the broadcast experience that many had acquired on radio. Most important, however, stage comics could maximize the visual immediacy of television. The National Director of TV Programming at NBC, Norman Blackburn, was quoted in *Variety* in early 1949: "Performers with stage and vaudeville background have taken to TV like a duck to water; they usually have a quick study, know what to do with their hands, and realize that a bit of sight business conveys much more meaning than the spoken word."[3] Talent agencies also were cognizant of the importance of the vaudeville comic and by 1948 were "shifting vaude specialists into the radio departments on the theory that it was more important to know the visual angle than to know broadcasting."[4]

To most critics, the predominance of comedians and lack of high-profile dramatic performers on television in the late 1940s signaled declining expectations for the medium as a "starmaker," whereas to many industry insiders it was only a matter of time until broadcasting would be able to create national comedic and dramatic stars of its own. What would eventually result, though, was that comics such as Milton Berle (whom critics routinely recognized as the medium's only true star from 1948 to 1949) would come to represent the particular aesthetic and commercial nature of television to such a degree as to redefine expectations for television's "star system." Comedy would prevail and television would continue to borrow and recast stars, genres, and performance styles from other entertainment industries.[5] And, for network television's first few years, the medium's stars would largely be hosts of variety programs. The influences of Broadway, nightclub revues, and vaudeville were vital to the delineation of effective performance models for television as they all utilized liveness, direct audience engagement, broad physicality, fast pacing, verbal dexterity, and common, deeply rooted cultural and historical references. Television reimagined many traditional forms of popular culture, adapting them to suit its particular needs and strengths. This process reveals not only the history of specific models of television performers and performance styles but also the development of genres, narrative strategies, production techniques, and economic relationships that, in one form or another, continue in television to this day.

Many had predicted that it would be the performers who could emphasize the visual nature of television who would end up being the medium's new stars. During 1946–1947, discussion in the trade papers focused on the potential that television presented for people in legitimate theater. Because live television required similar acting techniques and could be blocked according to a proscenium style, it was commonly thought to be akin to theater. The Broadway producer John Golden was quoted as saying, "Radio has been getting away with murder. All you've had to do on radio was to read. But reading isn't acting. Acting in television will be like that in the theater, an actor creating a role every night, developing it." He was also quick to add that television "is a blood-brother of the theater, closer to it, more akin than films or radio. It's as good as legit; it's the same creative art."[6] Lawrence Langner, co-director of the Theatre Guild, also perceived a similar link between the two entertainment media when he said that television acting "requires study, adequate rehearsal and as much thought and care as in a Broadway play."[7] Although the live anthology dramas of the mid-1950s would prove Golden and Langner to be right in many ways, in general, the norms of television performance would turn out to be quite different.

Although some theater producers hoped that television would be a well-paid, productive, and culturally significant outlet for their talent, some Hollywood producers looked at the new medium with disdain. The Motion Picture Producers Association (MPPA) enacted a ban on video appearances for their contract players during the first few years of the 1950s. But, as *Time* reported in the fall of 1953 (when the official ban was supposedly over), "most term contracts at the big cinema studios still forbid TV appearances, except for special walk-ons to plug a new picture."[8] Moreover, performers themselves were fairly hesitant about performing on the nascent medium. The broadcast industry tried to lure top name talent by increasing television fees and salaries by 250 to 500 percent during 1947.[9] Although many Broadway and nightclub performers flocked to broadcasting as a result, television would have to prove itself as a legitimate "starmaker" before most Hollywood stars would agree to appear on it in anything more than short guest appearances.[10]

In part because of the reluctance of Hollywood stars and the desire for format continuity on behalf of sponsors, the talent lineup for early television would contain many familiar radio names. It also included a number of vaudevillians who failed on radio and in film but eventually found success in television. Yet, as the managing director of the Roxy theater, A. J. Balaban, warned, "New talent and new faces, splendid though they might be, will still have many bitter lessons to learn in serving up the public's entertainment. While they're floundering, the experienced old-timers will swing easily into action and take a firm grip upon the affection of the new video audience."[11]

"When Vaudeville Died, Television was the Box that They Put it In"

He was a radio comic
and she, the girl of his choice
when she asked what he'd like for his birthday
he said in his radio voice
get me a coaxial cable
is that too much to ask?
I've been working over a hot mike for years
and now it's become such a task.
My rating is dropping by daily
what does Hooper want from me?
I haven't had a call from Paley
and it's so lonesome at NBC
so get me a coaxial cable
come on, be a good little girl
I'd like to televise my stuff myself
before it's done better by Berle

—Eddie Cantor[12]

One of the obvious types of performers expected to do well in television was the radio comic who also had experience in nightclubs, vaudeville, or on Broadway. Their particular combination of competencies, along with their familiarity among broadcast audiences, made these comics obvious candidates for television stardom. They also were being lured over to television during a time when resources and audiences were being drawn away from radio and toward the new medium. Because of the rise of television programming, competition from quiz and dramatic programs on radio, as well as complaints about the staid nature of gag comedy, many well-known radio comedians saw their ratings plummet in the late 1940s, which led them to ask Cantor's question, "What does Hooper want from me?" At times, it would seem that Hooper (or at least the audience measured by Hooper) wanted fresh talent and formats. As George Rosen reported in a 1949 front page story in *Variety*, "The era when a Jack Benny, Edgar Bergen or Bing Crosby was assured of a loyal following and an automatic top Hooper simply because they were a Benny, Bergen, or Crosby, no longer exists."[13] Nevertheless, sponsors and network heads were mindful of continuity and consistency in their approach to the impending arrival of television. They remained committed to bringing established radio talent to television.

Many radio comics, by contrast, remained wary of the new medium and worried about how they might fare if they chose to appear on it. They wanted to make the jump, but the transition appeared far from smooth. At the time, Norman Blackburn stated publicly that he believed "top name talent avoided TV for three reasons—one, they were waiting for it to grow up—two, they didn't need it—three, they were afraid of it."[14] What seemed to frighten radio comedians most was television's visual nature and its voracious appetite for material. It was assumed that radio work was far easier than television work because television required an incredible amount of stamina, memorization, and rehearsal from its talent. Unlike radio, where actors could read directly from scripts during broadcasts, television performers had to memorize scripts as well as act with their entire bodies not just their voices. Blackburn, in fact, went so far as to warn radio performers that if they chose to enter into television they would find that "no more is it possible to walk into the studio in street clothes or loafer jackets, give the script a fast reading, and then hurry over to the nearest pub to have a couple of quick ones before strolling leisurely back to look at the script cuts . . . [TV] entails hours of hard work and study."[15]

If the labor involved in television didn't bring about anxiety in the radio comic, its dependence on visual humor probably did. Although they most likely had prior stage experience, many had become used to the performance styles they had honed on radio. For these comics, performing on television would involve a return to old, perhaps rusty, techniques. As Bob Hope humorously remarked in 1947, "The advent of television will work profound changes in the field of radio comedy with performers finding it necessary to develop their humor from the visual rather than the aural branches of wit . . . Comedians will have available many of the old props that used to help punch a line. Fortunately, I still retain the electric bow tie that kept my act moving at Loews Pitkin."[16] Others had a different sort of anxiety regarding television's visuality: they worried about their attractiveness, fearing that audiences would perhaps finally notice their age, or some unattractive feature they longed to hide. A poem in a July 1947 issue of *Variety* gleefully addressed this issue outright: "We'll see, as well as hear, the gags: Benny's bald spot and Allen's bags [. . .]"[17]

In his book on radio comedy, Arthur Frank Wertheim argues that Fred Allen's failure in television can be at least partially explained by his physical imperfections, as television highlighted his baggy eyes, protruding jowls, and tendency to perspire. Wertheim also notes that Allen's ill health had affected his productivity and that his social satire just did not fit with the conservative attitudes of the 1950s.[18] Allen, along with Crosby, Jack Benny, Edgar Bergen, Groucho Marx, Eddie Cantor, and Burns, didn't enter into television until the 1950–1951 season.[19] Even though he had the top-ranked radio program on the air in 1946–1947, by 1949 Allen's ratings had dropped so low (mainly because of competition from ABC's *Name That Tune* with Bert Parks) that he left radio altogether in 1949. Allen initially appeared on NBC as one of the hosts of *The Colgate Comedy Hour* (1950–1955) and, after disparaging the format for years on his radio program, as a regular panelist on the quiz show *What's My Line* (1950–1967, CBS). By 1952 he was the host of two other quiz shows, *Two for the Money* (1952–1953, NBC; 1953–1957, CBS) and *Judge For Yourself* (1953–1954, NBC), as well as a rotating host (with Hope and Jerry Lester) of *Sound-Off Time* (1951–1952, NBC).

But Allen never acquired a significant following on television. In a 1950 review of one of his first video performances, Jack Gould wrote that Allen seemed bored and that he let his "disdain for the medium" carry through his entire performance, leading the reviewer to remark that he "did not seem to be trying very hard, which is not the trouper's way of showing loyalty to an audience that wants and expects to be entertained."[20] The media critic for the *Hollywood Citizen-News*, Robert C. Ruark, wrote that "it is easy to understand why Fred has made no real impact in the TV field, for his humor has always been off-center, and never embraced the pratfall or spitting-on people, hurling ice-cream or paint pots techniques that seem to be so popular these days."[21] Others noted that audiences were not very accepting of his attempts early on to bring the characters of *Allen's Alley* to television (he first tried to do it through puppets on *Colgate Comedy Hour* and then, briefly, with members of the original cast), perhaps because their representations were so different from what radio listeners had been imagining in their own minds all those years listening to radio. Then, when he moved into giveaways, Allen seemed ill-suited for the format, as it gave him little room to ad-lib or show off his verbal dexterity. His performances led Pat Weaver to remark later that "it broke my heart to watch him on TV."[22] Not surprisingly, Allen was one of the most outspoken critics of television and the lengths a comedian would have had to go to succeed in it. In an interview with *Life* in 1949, Allen complained:

Fig. 2.1   Fred Allen. Library of American Broadcasting.

The screen isn't the only small thing in television. Smallness seems to be the outstanding characteristic of the whole medium right now. It has small minds, small talents, small budgets. In fact you can take anything connected with television, and you'll find it so small that you can hide it in a flea's navel and still have enough room beside it for the heart of a network vice president.[23]

Fig. 2.2   All of the characters in *Allen's Alley* are on display in this 1948 ad for Ford.

Even before he started in the field, Allen believed that television would be an uphill battle for him and others like him, stating that, "we all have a great problem— Benny, Hope, all of us. We don't know how to duplicate our success in radio. … We don't know what will be funny or even whether our looks will be acceptable."[24] Yet many of the great radio comics *did* succeed in television. Some even blossomed. In reviewing NBC's

Fig. 2.3 Bob Hope's early television appearances were well received by critics. Library of American Broadcasting.

Star-Spangled Revue, Bob Hope's television debut, Gould wrote: "Bob himself was in rare form and a far cry from the talkative and repetitive Hope of radio renown. Given the freedom of movement, which is pure elixir to the comedian who for many years has been hobbled to a radio microphone, he had all the bounce and energy of a tot turned loose from a play pen."[25] Burns and Allen, Eddie Cantor, and Jack Benny also were praised by critics for their ability to bring the best of the verbal and the visual together in one package for television.

Essential to any performer's success in television was the ability to work with and highlight the medium's defining characteristics. William Boddy points out that television provided "a unique synthesis of the immediacy of the live theatrical performance, the space-conquering powers of radio, and the visual strategies of the motion picture."[26] And, as many television historians have noted, early discussions of the medium usually posited immediacy, liveness, and intimacy as the attributes that made television such a compelling visual medium. A widely held belief among industry insiders at this time was that vaudeville comedians were the performers whose talents could best exploit these "unique" characteristics. In fact, as early as 1936, entertainment reporters, such as Carroll Nye of the *Los Angeles Times*, were predicting that television would "bring back the vaude-villians who were shunted into obscurity with the advent of the talkies."[27]

Obviously, one of the main reasons television was assumed to be more intimate than film or stage was because of its position in the home or in smaller communal environments. Although radio also had provided such intimacy in regards to reception, television's visuality made it an even more tangible and powerful presence in home life. Popular rhetoric of the period described the experience of television as one that brought the world into the home and was often termed by the popular press as "home theater" or "family theater." However, the anxieties of consumer and social critics surrounding the deleterious impact television had on children and the more general fears of surveillance associated with the sets' placement in the private sphere provided the dark side of the technology's intimacy. Despite such worries, broadcasters understood that television's pertinence and influence with audiences would bode well for advertisers. Vaudeville's presentational mode, in particular, underscored the viewer's familiar relationship with television. On stage, a vaudeville performer courted his or her audience with direct address, colloquialisms, familiar ethnic tropes, and references to local habits, knowledge, politics, or places.[28] He or she would alter an act to fit a particular audience's interest, ethnic makeup, or responses to particular jokes. Audience interaction was essential to a comedian's performance and construction of character identity. When these performers moved into radio, producers quickly learned that they performed best when in front of a studio audience.[29] This was doubly true for the vaude acts that performed on television. Leo Bogart claimed that this use of the presentational mode to create a sense of intimacy with the audience when combined with star power led to a program's success: "the quality of direct and intimate contact . . . is carefully nurtured by skilled performers. This very illusion of personal communication with a glamorous, famous personality gives the broadcast media much of their appeal."[30] This strategy would cross over to television's early use of vaudeville performers in variety or variety/sit-com blends. *The Burns and Allen Show* (1950–1958, CBS) and *The Jack Benny Program* (1950–1964, CBS; 1964–1965, NBC) utilized the presentational mode as they brought situational context to their narratives both in radio and in television. Still, television insiders recognized that the vaudeville performance style could not be adopted wholesale from the stage. The inventor Lee De Forest predicted that "acting in television will assume a completely new method, different from that of the stage or radio. Since all of the audience is literally in the front row, there will be no necessity for throwing voices and emotions to the back of the theater, and the acting can become natural and unexaggerated."[31]

Television would not, as De Forest expected, require a complete rewriting of prior performance style. In fact, the industry desired its performers to utilize many stage techniques—both legitimate and vaudeville styles—in order to flaunt the visuality of the new medium. Even as William Eddy pointed out, "Vaudeville and night-club acts will generally require less readjustment for television than the straight stage and radio productions," he also recognized that the broad nature of such performance still had to be toned down and cautioned that "exaggerated action, together with other types of overplaying, will appear doubly distasteful when observed from the fireside seat in the familiar atmosphere of the home."[32] Thomas Hutchinson, one of the first to teach a class in television production at New York University, argued that the close-up in television would cause vaudevillians to be under such "minute scrutiny" they would be forced to relearn particular stage techniques. Because vaudevillians would no longer "need to project to the last row of the gallery," they would have to assume subtler gestures, expressions, and tones. Hutchinson, also a former television producer, recounts one of his own experiences with an audition in an effort to illustrate the problems that can arise when vaudeville acts don't alter their style for the domestic viewer:

> One team with an international reputation was suggested as television program material . . . they were both so used to projecting their material, to putting their personalities across the foot lights, that their television personality was far from what it had been in the theater. We saw them "work." We saw "the wheels go round." They strove so hard to put the act "over" that they were totally unsuited for the new medium.[33]

According to such criticism, the vaudevillian would need to accentuate some characteristics of the performance style while reigning in others. Most important, the brashness of vaudeville performance had to be tempered while the presentational mode of address was emphasized. Once these adjustments were instituted within the vaudevillian persona and "overplaying" and "distastefulness" were contained, the performers' penchant for visual humor, physical play, and intimate address would enable them to capture both the novelty of intimacy and the visuality that television was said to embody.

Beyond the characteristic of intimacy, however, the vaude performer was also considered to be well suited to the immediacy and spontaneity of the medium. In the early years, networks preferred live programming over filmed because it differentiated their product from that of Hollywood and underscored the necessity of network connection for affiliates. With regard to the latter, networks made their programs and their stars limited commodities by transmitting their fare live to their affiliate chain only at specific times.[34] Regarding the former, television's theatricality encapsulated the live and unpredictable potentialities of the medium's aesthetic nature. (As a result, another genre that dominated prime time in early television besides the variety show was anthology drama.) One of the elements of live television work that made it both taxing and exciting was that, unlike film actors who performed according to the boundaries and timing of the continuity script, television performers—whether dramatic or comedic—were expected to perform without retakes or breaks and with very little rehearsal time. This had the effect of not only recreating the theatrical experience for home viewers but also keeping them in a state of expectation, feeling as though anything could happen—good or bad—at any moment while they were watching the program. Certainly, this effect underscored the sense of immediacy or liveness of the moment, but also created spontaneity for the medium. Ad-libbing was one way in which performers played on this sense of spontaneity. It also had the added benefit of being a handy tool to cover mistakes or forgotten lines. George Burns acknowledged that "The main point is that people think your show is 'live'—that's your big selling job on TV. There's always the possibility of someone flubbing a line or getting a laugh when least expected. That's what gives our type of show an extra bang and keeps the audience keyed up."[35] Commentators assumed that vaudeville or stage comedians were not only the best ad-libbers but could handle all the other pressures of live television—and, in fact, use those pressures to his advantage—better than anyone else. Those in the industry believed that taxing work on the vaudeville or nightclub circuit was more than adequate preparation for the demands of television.

The variety programs, which were more popular with sponsors than were dramas, were staged in proscenium style, broadcast in front of a studio audience, and replicated much of the feel of a traditional vaudeville show. Producers wished to achieve a sense of transportation for their audience, to allow them to participate in a public entertainment event yet within the safety and familiarity of their own home. This view from the home was a privileged one, not only in its privacy but also in the fact that close-ups, framed views of the acts, and shots of the audience provided viewers with "the best seats in the house." Edward Stasheff notes that the variety format's intimate and theatrical presentation played a significant role in the genre's popularity. He writes that variety programs "possibly owe their success not only to their high-powered talent, but to the feeling they give the home viewer of having a front row seat among the members of a theater audience at a Broadway show. That's a good feeling to have in Hinterland, Iowa, or Suburbia, New Jersey."[36]

Yet the relationships between a performer, the studio audience, and home viewers were rather tricky, and they had to be experimented with before producers and comics found the right balance between creating intimacy and replicating the shared theatrical experience. Early on, networks and agencies planned on using large movie palaces and theater spaces, which could seat one thousand to two thousand people.[37] Although most comics genuinely wanted to do their shows in front of a live audience, this size audience seemed too large to some. They feared that it might cause them to "disconnect" from their television audiences and would likely "kill the very factor that has been most responsible for TVs rapid rise—its intimacy."[38] Cantor, one of the most outspoken critics of the move toward larger theaters, suggested that "the presence of large, enthusiastic, beyond-the-footlights crowds too frequently causes talent to lose sight of the fact that TV is basically home entertainment," adding, "Why should you jeopardize the show for 20,000,000 viewers for the benefit of 1,000 or so in the studio?"[39] Although Milton Berle insisted that his show originate from NBC's studio 6B, which would only accommodate 150 spectators, he, too, had to adjust to the presence of a camera and the implied presence of a home audience. A *Variety* reviewer recognized *Texaco Star Theatre*'s privileging of the studio audience in one episode saying that he

[had] the feeling that all of the performers, particularly Berle, were working for the benefit of the lens audience rather than the live

audience. That's a difficult thing to analyze and it won't be attempted, but in the past the feeling has been that the cameras were giving home-viewers a sneak look at shows stage for those in attendance. This show sponsored a reverse thought.[40]

Later, Gould and others would praise Berle for his singular talent for creating an intimate, spontaneous, and natural bond with the home viewer, while not losing the attention or enthusiastic response of the studio audience. Other comics would learn from Berle and come to find their own way to negotiate their two audiences with varying degrees of success.[41]

Besides learning how to play to two very different audiences, vaudeville comics had to tweak other aspects of their performance style in order to survive on the new medium. Very early experimental variety programs provided networks, producers, and talent with the opportunity to try out various strategies to better translate the genre from stage to television.[42] One such program was Standard Brands's variety show *Hour Glass*, which began airing on NBC in May 1946. The first few episodes consisted of comedy sketches, musical numbers, dance acts, and, on one occasion, a film of South American dancing. These early episodes were considered flops, with many critics finding little or nothing to praise about the show. One *Variety* reviewer complained that the producers didn't seem to know what to do with their talent as "the stars were apparently just set out in front of the camera and told to do their stuff."[43] *Hour Glass*'s producer Edward Sobol (who had worked as a vaudeville stage manager and director prior to coming to television) found it difficult to properly capture the physicality of many of the routines because of the limitations of television production facilities—the stages were too small, the cameras didn't capture enough detail, and the lighting was severe—saying that since "[vaudeville] acts are accustomed to the freedom of movement allowed by the stage or night club, it is difficult to restrict them to the television playing areas."[44] Moreover, the comedians' pacing seemed off for the "inherent intimate aspects of television," so Sobol recommended that they should not only quicken their pace, but "should be limited to three minutes [of airtime] a piece."[45] By November of that year, *Hour Glass* had worked out many of its kinks, and producers had added a radio star, Edgar Bergen, as the program's host.[46] Reviews were much improved this time around, with *Variety* noting that since it was "the first time any of the top 15 Hooperated radio stars was featured in a video production," the show "proved, if anything, that a good radio comedian is equally good on television even without the aid of script…if other radio and screen stars follow his lead now, television might get that needed stimulus."[47] Judine Mayerle argues that the *Hour Glass* provided the model for the variety shows that followed and showed that "with a few alterations, a simple vaudeville format could be tailor-made for the screen."[48] As Mayerle's research reveals, NBC developed other variety programs from 1946 to 1947 to experiment with various techniques and material. In addition, during that same time, John Royal and production chief Warren Wade came up with a unique way to deal with the challenge of training and developing on screen talent. In many ways, their plan mimicked an old-fashioned vaudeville circuit as it depended on local stations casting and training their own talent for their own repertory groups. They would then have each repertory group travel from station to station performing their one "episode" or "play." Local stations would benefit from having a range of talent and programming, as performers and crew gained invaluable experience perfecting their act or craft.[49]

By 1948, as television was exiting its experimental stage and entering into full commercial operation, one variety program became a national phenomenon and solidified the viability of the televised variety format—*Texaco Star Theatre* (1948–1952, NBC). At the time of its premiere, NBC had seven stations as a part of its network in New York, Washington, Philadelphia, Schenectady, Baltimore, Boston, and Chicago. The network would add Los Angeles to its roster as well by the end of the year and David Sarnoff was estimating that there were more than three hundred thousand television sets in use in the United States.[50] Even though NBC had already aired vaudeville type programs, Texaco took out an ad on the front page of *Variety* on May 19, 1948, announcing: "A ghost out of the past and a show biz potential of the future—vaudeville and television—will be mated June 6 when *Texaco Star Theatre* stars a series of vaude programs."[51] (This mating of vaudeville and video at that point led to the use of a new term in the industry: vaudeo.) Berle, the program's host, quickly became known as an exemplary variety performer who was able to sustain an extremely high level of energy during a taxing production schedule. Those who watched Berle perform and those who worked with him frequently commented on his frenetic energy and ability to go without sleep. The writer Edwin James noted that "Berle, who cannot abide idleness gets into almost every act. He sings with singers, dances with dancers, tells jokes with other comedians, and tumbles with acrobats. An ordinary mortal would collapse in the midst of such frenzy.

It only stimulates Berle."[52] His performance style embodied the characteristics of immediacy, spontaneity, intimacy; was often credited with selling "a million television sets;" and it was described by critics as a "whirling dervish,"[53] an "inexhaustible package marked explosive,"[54] and "a dozen or more men rolled into one."[55] He had, quite literally, come to represent the television medium in the late 1940s and early 1950s and, consequently, the mythology of his initial popularity with audiences rivaled that of radio's *Amos 'n Andy*. Not surprisingly, *Variety* listed him as the highlight of television programming in 1948–1949, calling him (as many would) "Mr. Television." *Variety* explained that "When, single-handedly, you can drive the taxis off the streets of New York between 8 and 9 on a Tuesday night; reconstruct neighborhood patterns so that stores shut down Tuesday nights instead of Wednesdays, and inject a showmanship into programming so that video could compete favorably with the more established show biz media—then you rate the accolade of 'Mr. Television' of 1949."[56] Bob Considine of *The New York Journal-American* went so far as to claim that "TV gets better every day. Yet we wonder if its bright future will see it produce a figure quite as overpowering as Milton Berle. He made television, just as surely as Chaplin and Pickford made the movie industry."[57] Berle's aggressive emphasis on the physical aspects of comedy, his slick vaudeville routines, expressive gestures, and quick tongue enabled him to succeed in an industry that was looking to highlight its visuality and immediacy.

In many ways, Berle also was television's consummate intertextual performer, as he not only followed the vaudeo star's familiar path from stage to radio to film to television, but also was a poacher of some of the best comic material from all of those entertainment forms. A highly successful vaudeville and nightclub performer throughout the 1920s and 1930s, Berle had difficulty manipulating his persona to fit the needs of other entertainment forms. He seemed to be at his best when performing unrestrained in a visual environment in front of an audience. "Berle's success on television is a curious byproduct of repeated flops in both radio and movies," reported *Time* in 1951. "The flops hurt deeply and worried him about his appeal to a mass audience. But they forced him into well-paid jobs in nightclubs, where live audiences kept his talents supple. Meanwhile, more successful comedians were falling into the lazier habit of peering at scripts through spectacles."[58]

Berle, like a number of television stars who would follow, was heavily involved in the production of his own program, having a say in every detail

Fig. 2.4   Milton Berle was described by one critic as an "inexhaustible package marked 'explosive.'"
Library of American Broadcasting.

including the lighting, choreography, and costumes. Consequently, he
sustained considerable authority over his own image (at least, how it was
presented within the text of the program because his sponsor, Texaco, also

would make a considerable contribution to the refinement of the Berle persona as would NBC). During *Texaco Star Theatre*'s first year Berle was even the show's sole writer, culling from his vast joke file that contained hundreds of thousands of jokes, not all of them belonging to him.[59] Virtually every major comedian of the period complained that Berle had stolen material from them at one point or another in their careers. Hope told *Time* that "When you see Berle, you're seeing the best things of anybody who has ever been on Broadway. . . . I want to get into television before he uses up all my material."[60] Fred Allen joked that "[Berle's] a parrot with skin on."[61] Replying to those who accused him of stealing, Berle always asserted his belief that jokes are public property. This appropriation of material from other radio, vaudeville, and film stars resonated with the way that television recycled performers and genres from other entertainment media, ultimately underscoring its intertextuality.

Pat Weaver, the Variety Format, and Comedy Stars at NBC

Although *Texaco Star Theatre* and Berle proved vaudeo to be a successful genre for television, there were still aspects of the format that needed to be expanded on in order to keep it interesting over the long term. NBC was invested in being a leader in comedy during this period and put into effect a number of different comedy development plans beginning in 1949, which resulted from the hiring of Sylvester "Pat" Weaver as vice president in charge of television.

Not long after he took his new post, Weaver was speaking publicly about his drive to beat out CBS in the field of comedy and his desire to collect a roster of comedy "giants" and provide them with "great prestige and presentation" that would marry "spectacle and intimacy" as well as "continuity and freshness."[62] NBC had lost viewers and sponsors along with stars to CBS in the talent raid and Weaver was looking to bring back NBC's former first-place status with stars and high-quality comedy programs.[63] He also saw comedies as gateway programs that could attract a mass audience and then act as a lead-in to more serious fare. (Weaver worked to have the 8 P.M. programming slot become known as "comedy hour" on NBC, often placing dramas or prestige programs in the 9 P.M. time slot.) The model for his comedy programming and development plan would be the *Colgate Comedy Hour*, a big-budget variety program produced by the network and then sold to a single sponsor—Colgate. The program alternated big-name hosts such as Bob Hope, Eddie Cantor,

© FABIAN BACHRACH

Fig. 2.5   A young Pat Weaver. Library of American Broadcasting.

Abbott and Costello, Donald O'Connor, and Martin and Lewis and was scheduled for Saturdays at 8 P.M. (a time that Weaver wanted to become known as "comedy hour" all nights of the week on NBC). The concept of a rotating lineup of comics was intended to prevent their overexposure with audiences and their exhaustion from a weekly production schedule as well as prevent their writers from running through material too quickly.[64] Shows were built around the star that would be hosting that night. For example, the premiere episode of the program, which was handled like a Hollywood opening night with cameras showing a red carpet, searchlights, and celebrities entering the theater before the show began, showcased Cantor in many of his most beloved and familiar routines

DEAN MARTIN and JERRY LEWIS • Marion Marshall, Polly Bergen • Starring in the Hal Wallis Motion Picture, "THAT'S MY BOY" • A Paramount Release

Join the Stars with *Magnavox* **Big-Picture TV**

YOU join mighty proud company when you move magnificent Magnavox Big-Picture Television into your living room. For Magnavox instruments are built to grace the finest homes. And Magnavox Big-Picture TV is the kind enjoyed by so many of the entertainment world's most hard-to-please

experts. Owners tell us that the glorious, concert-hall tone of Magnavox and its noticeably clearer, sharper pictures are the envy of their neighborhoods. Magnavox combines advanced engineering—super-sensitive circuits, extra-powerful speakers and eye-restful filters—with stunning cabinetry of heirloom

quality. Yet Magnavox values are without equal. Choose the perfect Magnavox for your proud home at one of the distinguished dealers listed in your classified telephone directory. Only stores famous for outstanding service are selected to sell Magnavox. The Magnavox Company, Fort Wayne 4, Indiana.

the magnificent

THE FRENCH PROVINCIAL (also shown above), AM-FM radio-phonograph in smart Savoy finish. Accommodates twenty-inch TV now or later.

Magnavox
television - radio - phonograph

BETTER SIGHT...BETTER SOUND...BETTER BUY

Fig. 2.6   Magnavox asks viewers to "Join the Stars" (like Martin and Lewis) with their Big-Picture TV.

—including his songs "Banjo Eyes" and "He's Making Eyes at Me," which he sang in blackface. Moreover, programs would have separate production units (everything from producers to orchestras) for each star, so that Cantor would be working with a completely different set of people than, say, Bob Hope. Although one of the most expensive television programs ever aired up to that point, *Colgate Comedy Hour* proved to be a hit, cutting the ratings of its competition (*Toast of the Town*) in half on its first night on the air.[65] It also was a first step in Weaver's plan to get creative control out of the hands of advertising agencies and sponsors and into those of networks and talent agencies as it revealed the benefits of network production to both producers and sponsors. By year's end, Weaver was not only rotating comics on the program, he was also rotating advertisers.[66]

Two other programs resulted from Weaver's comedy plan in 1950, *All Star Revue* and *Your Show of Shows*. *All Star Revue* (which would later be renamed *Four-Star Revue*) was similar in concept to *Colgate Comedy Hour*, as it used a number of well-established comedy stars (Ed Wynn, Danny Thomas, Jack Carson, and Jimmy Durante) to host the variety program on a rotating basis, but it did not do well in the ratings as it was never a match for its competition—*Arthur Godfrey and Friends* on CBS. *Your Show of Shows* developed out of *Admiral Broadway Revue*, a program produced by Max Liebman and starring Sid Caesar and Imogene Coca, which was pulled off the air after nineteen weeks by its sponsor, despite the fact that it had won its time slot and was a hit with the critics.[67] *Your Show of Shows* was initially packaged as a part of Weaver's *Saturday Night Revue*—a two-and-a-half hour programming extravagance that also included an hour-long variety program hosted by Jack Carter.[68] *Your Show of Shows* was always described as being more Broadway revue than vaudeville show and was touted as a sophisticated and lavish production. Caesar performed sketches, developing characters, dialects, movie satires, and a style of pantomime that became famous for, most often in tandem with Coca. Interspersed between the segments that featured Coca and Caesar were orchestral and dance numbers, sometimes including opera or ballet pieces. It was, in many ways, the very smart, challenging, and yet entertaining program that Weaver had envisioned as a start to his "Operation Frontal Lobes" plan to slowly indoctrinate audiences to appreciate more "high" culture material. It also provided a new blueprint for production of comedy programs, as all the talent—including writers—remained the same (and retained full creative control) each week. It proved that individual writers would not simply run out of material if part of their job was to construct long-term sketches and characters for the comedians, not just one-off gags.

Weaver's plan resulted in NBC being the leader in comedy by the 1950–1951 season as *Texaco Star Theatre*, *Your Show of Shows,* and *The Colgate Comedy Hour* were all in the top five rated programs of the year. Over at CBS, however, Bill Paley did retain a small, but powerful, cadre of

Fig. 2.7   Sid Caesar and Imogene Coca. Library of American Broadcasting.

comedy stars and program hosts such as Jack Benny, Burns and Allen, Molly Goldberg, Ed Sullivan, and Arthur Godfrey, whereas Dumont had acquired Jackie Gleason for *Cavalcade of Stars*. (ABC had just begun to invest in comedy and was still relying heavily on films, music programs, wrestling, roller derbies, and a few action programs to make up their prime-time lineup.) It seems that both Paley and Weaver were correct in their assumption that comedy stars—former radio and stage performers alike—would be the key to their networks' early success with audiences. Those stars also would help them construct a network identity as they eventually became markers of a distinctive house style the longer they were associated with a particular network. Reporting in the spring of 1950 that CBS was continuing its emphasis on stars (a plan the publication had dubbed "Paley's Comet" six months prior), *Variety* quoted Paley as saying that a "network is as good as its comedy line-up."[69] Both CBS and NBC began to take out ads positioning themselves as one version or another of "network of the stars" or the "choice of America's most popular stars."[70] But there was more than just familiarity at work in the presold celebrity of many of the comedic performers that worked on television during these early years. Audiences also responded to them for specific cultural and socio-political reasons. Their on- and off-stage personas resonated with cultural memories as well as with values, norms, and meanings that were being negotiated at a time of great social change, which is why a number of them became more than just big-name performers; they became iconic. In order to more fully explicate the meaning of their personas and their popularity, the following chapter will focus less on the industrial drives that led to the rise of the vaudeo star in the late 1940s and early 1950s and focus more on the cultural ones.

CHAPTER 3
LESSONS FROM UNCLE MILTIE
Ethnic Masculinity and the Vaudeo Star

Many of the obituaries and tributes to Milton Berle published in the days following his death on March 27, 2002, emphasized the ways in which the comic's Jewishness either informed his comedy or was central to his stardom. In the *Wall Street Journal*, Joseph Epstein argued that "everyone had to know that Berle was Jewish," whereas the *Baltimore Sun*'s television critic detailed the ways that Berle's Jewishness was central to the rise and decline of his program *Texaco Star Theatre* and, more generally, the future representation of Jews on television.[1] Franklin Foer put it bluntly in his *Salon.com* article by simply (and rather lovingly) calling the comic a "very Jewy Jew."[2] Although it may be easy for contemporary critics to speak so frankly about Berle's ethnicity and its impact on his popularity, this type of commentary was not present in the mainstream coverage of the star during the years in which he was first dubbed "Mr. Television" and America's "Uncle Miltie." Rather, the signs of his Jewishness were read through his historical connections to things such as vaudeville, New York, and a particular type of ethnic masculinity.

After the proven commercial success of early variety programs, vaudeville-trained performers became the most sought-after personalities in television. Jack Benny signed an unprecedented ten-year contract with CBS worth almost $1 million in 1950, and the following year Berle

Fig. 3.1 Berle wears his Texaco uniform. (Author's personal collection.)

penned a deal with NBC that would cost the network up to $200,000 annually for thirty years. Clearly, networks and advertisers were confident about the long-term earning potential of such stars—in fact, it would seem as if they assumed the vaudeo trend in television would continue in perpetuity.[3] What actually occurred, however, is that vaudeo was absorbed into the sitcom format by the end of the decade and many of the genre's top stars watched their careers fizzle out by the early sixties. But during this early period, the variety performer's brash, stagy, New York vaudeville style became one of the most significant prototypes of the television star.

As the preceding chapter has shown, the broadcast industry was initially interested in exploiting ex-vaudeville stars for the way in which their performance style emphasized the visuality, spontaneity, immediacy, and intimacy of the television medium. Also, on a very practical level, ex-vaudevillians were a convenient pool of talent from which to draw, as many of them were New York regional performers and the television networks were all broadcasting from flagship stations in Manhattan. Yet, as a consequence of poaching vaudeville performance styles, the television industry was forced to confront the more indelicate aspects of variety-format humor that threatened to erupt unexpectedly on live television. Ex-vaudevillians had a tendency to ad-lib sly sight gags, asides, and subtle gestures to connote sexual references or situations. So, in this perilous terrain of a live, visual, domestic-entertainment medium, not only did the bawdy antics of stage comedy have to be eliminated from scripts in pre-production but also their spontaneous appearance in the program had to be anticipated. One way to ease the reception of such content was to contain or construct rhetorically a personality for the comedian that would befit the values of the television audience.

In relying on the proven variety format and using older vaudevillians and radio performers, the television industry was referencing a traditional form of entertainment while simultaneously promoting the novelty, the "newness," of the medium itself. It is in this context that the analysis of the appeal and cultural resonance of the early television comics becomes so significant. Considering the somewhat morally tenuous state of the variety format and the aging, ethnic, and somewhat flexible masculinity of the top comedy stars alongside the industry's desire to be perceived as a natural extension of the American family, the television industry might appear to have been working at cross-purposes.[4] However, these apparent contradictions in the construction of early comedy stars bespeak rather coherent symbolic constructions of ethnicity, masculinity, and anxieties over the changing demographics of the American cultural landscape.

Although a variety performer's particular ethnic identity was rarely addressed outright in television, traces of his ethnicity (usually his "Jewishness") were embedded in the star's persona and performance style. Specifically, the cultural and religious heritage of such individuals as Berle, George Burns, Eddie Cantor, Sid Caesar, and Jack Benny were obliquely referenced through their connection to the traditions of vaudeville, their affiliations with particular geographic areas/neighborhoods, their relationships with their extended families, and their representations of a historically and culturally located feminized masculinity. The subtle nature of these ethnic cues helped these comedians address the cultural experiences of the largely northern and urban television audience of the late 1940s without completely alienating viewers from other regions.

Comedy, Nostalgia, and Urban Ethnic Masculinity

Many vaudeo performers, including Berle (formerly Milton Berlinger), Caesar, Ed Wynn, Burns (Nathan Birnbaum), and Benny (Benjamin Kubelsky) had embarked (with varying degrees of success) on radio and film careers before their entry into television. Unlike people such as Benny and Burns, however, Berle was a risky investment for NBC, as his radio program *The Philip Morris Playhouse* and the radio version of *Texaco Star Theater* were largely considered disappointments.

Yet the physical comedy and sight gags Berle acquired during his years on the vaudeville circuit forecast a nice fit with the still-developing aesthetic economy of television. Given the collection of performers who were the early stars of television, it is a bit ironic that an industry so concerned with emphasizing the visual aspects of the new technology chose individuals so significantly older and less glamorous than their counterparts in film. As postwar Hollywood promoted the virile hero and the psychologically tormented detective, television offered a less traditional vision of maleness. In 1948, the year Berle became a household name with *Texaco Star Theatre*, films were released starring Montgomery Clift (*The Search*, *Red River*), John Wayne (*Red River*), Humphrey Bogart (*Key Largo*, *The Treasure of the Sierra Madre*), and Clark Gable (*Command Decision*) in roles as war heroes, prospectors, tough guys, and cowboys. Although film noir presented psychologically troubled and morally conflicted protagonists during these years, the presentation of masculinity through Hollywood product was still within the traditional white Anglo-Saxon masculine paradigm. Even Hollywood screen comedy had moved away from the anarchistic, vaudeville-inflected format and toward the narratively integrated romantic comedy that highlighted stars such as Gable and Cary Grant.

Compared to the more hardened male types found in film, the goofy, gangly, mugging Berle hardly appeared to be anyone's ideal man, or man-to-be. Nevertheless, his awkward physicality and psychological vulnerability expressed through self-deprecating humor were integral to the currency of his urban-based character. In his study of Berle's career, Arthur Frank

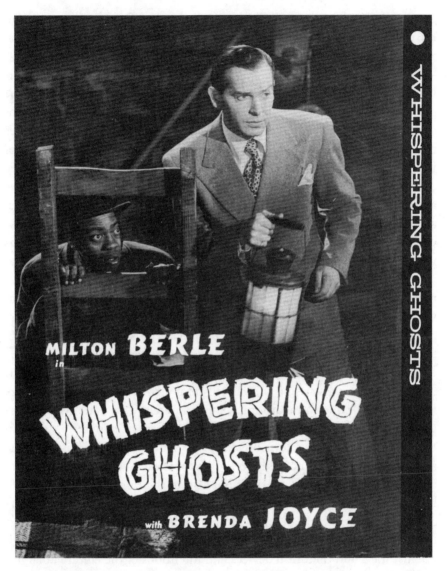

Fig. 3.2   Berle, who never made it big in Hollywood, starred in this 1942 film as a performer playing a detective on a radio program.

Wertheim found that Berle "personified a flippant city slicker—a character most viewers could understand."[5] Berle's character mobilized nostalgia for turn-of-the-century city culture, and more specifically for working-class theater (which depended on a revisionist history of its original "amoral" discursive construction). Although Berle certainly was not the last word in postwar virility, he possessed a homey familiarity and nostalgic resonance that was paramount to his popularity. And, surprisingly, his gender play and brash sensibilities actually worked for an audience whom the industry assumed was seeking moral reassurance and reinforcement of middle-class values from the new domestic medium.

Masculinity, in a strategic cultural performance of signifiers, is mobilized in relation to other social constructions and positions. In reading the personas of television stars of the late 1940s and early 1950s, the functions of ethnicity and class are often difficult to disentangle from gender codes because they work to inform one another. In discussing the process of identification, Judith Butler points out that individuals may wish to see coherence and authenticity in gender identity, but "[words, acts, gestures, and desire] are fabrications manufactured and sustained through corporeal signs and other discursive means."[6] A consequence of understanding masculinity as performance is that its mobilization as a congruency of signs within the system of media representation produces a mirroring effect that is overwhelming in its obfuscation. In television, as in film, performers represented and reconstituted signs present in contemporary culture that were meant to signal a specific form of gender identity. Thus, the performance of gender witnessed in everyday "reality" is replayed for viewers through these individuals' screen acts, words, and gestures and contributes to a further distancing of the signs of identity from its human object. Nevertheless, by discussing the various reference points each type of performance is aiming to incorporate, we can perceive an underlying structure informing individual characterizations of specific identity types. In the case of many early television stars, the "fabrications" of gender, class, and ethnicity are vital not only to their extratextual performance of functionally idealized public identities but also to their on-screen allusions to historically and generically defined characterizations. Their on- and off-screen personas, which are most often co-determined, are constituted via the text of the television shows in which they perform as well as through the historical and rhetorical devices and intentions of the culture at large.

Vaudeo stars were peculiarly adept at playing with signs of gender and ethnicity, as they were required by the variety format to inhabit numerous character types simultaneously. Sid Caesar, for example, was known for his portrayals of a regular cast of characters. When in his role as host of *Your Show of Shows*, he would perform a rather straight version of himself—or at least a consistent representation of who he was purported to be in the press. In his sketches, however, he created characters such as storyteller Somerset Winterset, the German professor, jazz musicians Progress Hornsby and Cool Cees, and, along with his partner, Imogene

Coca, Doris and Charlie Hickenlooper, an "average" middle-class Staten Island couple. It would seem that with Caesar's ability to take on so many characterizations, locating one stable identity for him would be difficult for the audience. Yet, as prolific as his representations of different identities were, his construction through the cultural legacies that arose out of the variety format, along with the publicity materials that accompanied his rise to television stardom, assembled a context for Caesar's reception as an individual and hinted at his own "authentic" gender and ethnic identity.

Most important to the creation of such a context was the origin of the vaudeo performance style. The variety format was understood as a modern extension of turn-of-the-century vaudeville, an industry populated largely by working-class performers of particular ethnic backgrounds, the most predominant being Irish and Jewish. Many vaudeo stars, including Caesar, had begun their careers in the vaudeville circuit and had later performed in the "Borscht Belt," a collection of Jewish hotels in the Catskill Mountains, further emphasizing the performers' ethnic and religious affiliations.

The publicity generated for Caesar, Berle, and Burns consistently referred to their early careers in Borscht Belt venues, as well as to their childhood and familial connections in such New York ethnic neighborhoods as Harlem, Yonkers, and the Lower East Side. In some instances, articles would reveal not only the neighborhood of a performer's birth but also the exact street address.[7] These elements formed the basis for these performers' reception as men who maintained affiliations with specific regional and cultural conceptions of what it meant to be a man. They were clearly not raised to become traditional Anglo-Saxon heroes in the form of cowboys, drifters, or tough guys. Instead, their roles as comedic performers were constructed and received through historically and culturally specific signifiers of ethnic masculinity. In addition, the variety format in which they worked (with its emphasis on multiple characters, ethnic humor, and drag) granted them the room in which to play with these signifiers and further complicating assumptions about stable, traditional American masculinity.

Many of the stock characters comedians such as Berle and Benny created for their variety programs were amalgams of American assumptions about masculinity and ethnicity. The implementation of easily recognizable class, ethnic, gender, and regional traits into vaudeville characters allowed for the lean economy of the form's truncated narratives and "olio" structure. Jewish characters, which did not become popular in vaudeville until the early 1900s, often were created through the donning of a long, pointed beard, large spectacles, a long black coat, and a dark plug hat.[8] Michele Hilmes notes that this vaudevillian practice continued in Golden Age radio, but because it was required to reside in the purely aural, the skills of the ethnic dialectician were emphasized.[9] Race, ethnicity, and class were fashioned through vocal cues and underscored by stereotyped behaviors. This practice of turning performance into a shorthand for identity was picked up in vaudeo, allowing a similar economy to inform variety-show production. Although dialect continued to play a role in broadcast variety programs, there also was a reemphasis on visual display, as performers could again rely on mannerisms and dress to connote identity.

Although the vaudeo star's own ethnicity did not always dovetail with the ethnicity of the characters he or she created, the audience may have assumed a certain amount of cross-fertilization between performers' on-stage characters and their authentic identities.[10] It was an assumption that would have been furthered by the use of the vaudeville aesthetic, because it overtly courted the conflation of on-stage/off-stage personas by breaking the boundaries of theatrical realism, making possible a higher degree of intimacy between performer and audience. Variety programs borrowed vaudeville style and continued to emphasize the relationship between performer and audiences through direct address and studio-audience interaction. Television performers were also required to act sincere and trustworthy in order to sell their sponsors' wares, they often worked to come across as more genuine. For vaudeo stars, the representation of what could be construed as their authentic on- and off-stage personas occurred while they were acting as their show's host—during monologues and in their (relatively) straight interactions with their guest stars. They would then go on to perform in sketches as a number of different characters, but, in self-referential asides to their audience, they would often return to what might be understood to be their core personality.

The identities of early television comedians interacted with and were influenced by both the operations of television stardom and those of comedy performance. Moreover, the ethnicity and sexuality of these stars were at stake both in their on-screen performance of specific character types and in the construction of their off-screen personas. As such, the disruptive nature of what Steven Seidman calls "comedian-centered comedy" has ramifications for the audience's subsequent perceptions of the comedian's "authentic" identity in relation to his or her character-izations.[11] Work on the comedy genre by Seidman, Henry Jenkins, and

Frank Krutnik, among others, reveals the ways in which comedian-centered gag or slapstick comedy often disrupts classical narrative continuity by enabling the audience to linger on the spectacle of the performance. In discussions of film comedy, examples such as Hope's cross-dressing in *Road to Rio* (Norman Z. McLeod, 1947) are used to elucidate the way in which the comedian's over-the-top performance can intrude into an otherwise coherent, stable narrative.

Scholars also suggest that such narrative ruptures may destabilize a comedy star's characterization/identity. Speaking specifically about classical Hollywood comedy, Krutnik argues that "the comedian figure deforms familiar conventions of film heroism, unified identity, and mature sexuality," concluding that, in the play of narrative disruption and containment, these films "circulate around questions of gendered identity."[12] The variety format is not as coherent in its structure as the classical Hollywood film. However, the vaudeo star's ability to cross over from one sketch to another, to embody numerous characters, to provide a base or stable personality in his role as host, and spontaneously to interrupt guest performances reveals similar moments of disjuncture caused by the seemingly uncontrolled nature of the lead performer.[13] Vaudeo's presentational, comedian-centered, gag- and slapstick-style figure the vaudeo comic's persona as one that is fluid in its relation both to narrative and to constructions of authenticity and performance. Specifically, vaudeo appears to be fascinated with the reticulations of gender and ethnicity. The vaudeo star plays with the signs of both these categories of identity and complicates them through his decentered position within the narrative and his intimate relationship with his audience.

Berle's performance style provides a good example of this process, as he was known for constantly interrupting his show's sketches and guest stars' performances. While a sketch was in progress, Berle would make asides to his audiences out of the sketch's character, often references or inside jokes to what the audience knew of his on-stage persona. When performing his drag bits, Berle's extreme physical plays on femininity (such as batting his false eyelashes or pursing his overly made-up lips in a faux-seductive manner) were often disrupted by moments when he would lower his voice and speak to the audience in his normal baritone. For example, in one 1951 episode of *Texaco Star Theatre*, Berle opened the show as a bride escorted by a portly groom. The audience howled as he swished his hips and batted his eyes, as he sang a ditty about how he married an elephant. As the groom left the stage, Berle continued his monologue in his own voice and as Berle, but with the occasional feminine gesture. Still dressed as a woman, Berle discussed the confinements and conflicts of marriage from a male point of view, commiserating with the married men in the audience. He ended by taking off his wig and introducing the evening's guest star. Moments like this served to disrupt the continuity of Berle's characterizations, highlighting their artificiality and confusing the distinctions of gender codes. Berle also "spontaneously" broke into the acts performed by his musical guests. The mini-narratives of even these smaller moments within the variety format were made discontinuous by Berle's intrusive persona.

Although Berle is an extreme example of the decenteredness of the vaudeo star, Burns and Caesar also went in and out of character. In fact, Burns's oft-discussed movement across genres as the simultaneous host and sitcom star of *The Burns and Allen Show* (1950–1958, CBS) is an excellent example of the way such a performer could enter and exit a self-contained narrative. This process resulted in a destabilization of all characterization, emphasizing the constructed nature of identity, but at the same time it reasserted the relatively consistent personality that existed beneath the performances of these stars.

Jewish Identity and the Vaudeo Star

The strong conflation of "real life" and "reel life" that occurred in the reception of the television comedian was contained within the destabilized identity that is part of the narrative play in comedian-centered comedy. Thus, the comedian's persona may have been rendered as contradictory or even incoherent. In attempting to forge a unified identity for these performers that satisfied the perceived desires of the domestic broadcast audience, the network publicity offices and talent agencies had to assemble a narrative for the lives of the stars that assimilated the various, and often disjunctive, aspects of the comedians' sexuality and ethnicity. This was most often achieved through discourses on the "real-life" ethnic and familial pasts of vaudeo performers. Many of the men who were heralded as the first comedians to consolidate a regular viewing audience were fashioned as ethically (and ethnically) justified recipients of television fame. This meant that discourses on domesticity were constructed around these men that spoke to pertinent social concerns and memories that were circulating in the postwar culture.

Discourses that served to remind viewers of the ethnic and immigrant makeup of the urban areas from which performers came were particularly important elements in positioning the domestic histories of the typical vaudeo star. However, the very ethnic identities and working-class origins of these vaudeo stars that were comforting reminders of America's immigrant past also could be deleterious elements of their personas in the years immediately following World War II. Therefore, few stars openly acknowledged their ethnicity—especially those who were Jewish, except for occasionally noting their birth name (i.e., that Eddie Cantor was born Edward Israel Iskowitz), the use of Yiddish words such as *tuchus* or *kishkas* or, in the case of George Burns and Gracie Allen, acknowledging a "mixed" marriage. Instead, covert signs of ethnic identity were embedded in the representations of the performers' work and familial histories, their constructions of feminized masculinities, and their imbrication in urban values.

Barry Rubin asserts that in the immediate postwar era relatively few Jews living in the public eye would play openly on their backgrounds, as "for most, the market's dictate did more than any prejudice to make it preferable not to seem too Jewish lest this cut one off from the best opportunities and widest audience."[14] Rubin remarks that, although comedy was dominated by Jews and Jewish humor, "the Jewish comedians, all products of Orthodox, Yiddish-speaking homes, retained none of these characteristics themselves." Instead, audiences could infer a performer's ethnic background through the stories told about his past as well as subtle inflections, gestures, and regional references.[15]

Rubin notes that Americans in the 1950s developed a fascination with Jewish culture because it represented difference in an era of extreme homogeneity. He quotes Robert Alter, who wrote frequently on Jewish identity, as saying that it was significant during this period that "the Jew has a special language, a unique system of gestures, a different kind of history which goes much further back than that of other Americans, a different cuisine, a kind of humor and irony that other Americans don't have, the colorfulness and pathos which other Americans aren't supposed to possess anymore."[16] These differences acted as nostalgic reminders of the diverse nationalities and cultures that had been assimilated into the country's urban culture during the waves of immigration in the nineteenth century.

Although the label of immigrant was not synonymous with a particular ethnicity, the influx of immigrant groups into poor and working-class neighborhoods of New York at the turn of the century led to the confluence of an individual's ethnic background with the urban space in which he or she was raised. For example, much of Burns's extratextual material mentioned his years growing up as the child of immigrants on New York's Lower East Side. Although not every resident of this area was of Jewish extraction, it was historically known by East Coast Americans as a largely Jewish enclave: a first stop for Eastern European Jews in their initiation into U.S. culture. Thus, the performance and speaking style George Burns acquired from the Lower East Side were clues to his ethnic and religious background. Sid Caesar gave similar clues to his background. Karen Adair suggests that "though Caesar never made a point of his Jewishness while performing, nevertheless it did emerge in some of his inflections and phrasings, such as 'Darts, they're playing.' The almost Talmudic lament of 'It'll be a miracle!' is the sort of thing Jewish mothers wail when their kids are growing up."[17]

Jenkins's study of Cantor's film persona reveals the ways in which the comedian's Jewishness was assimilated into general connotations of urbanity in the late 1920s and early 1930s as a way to appeal to film audiences in regions outside of East Coast urban centers.[18] As a result, Cantor (who had made much of his Jewishness in publicity materials early in his career) was forced to avoid direct references to his ethnicity and instead allowed his persona to retain only "subtle textual traces of his Jewishness, all but invisible to regional viewers, yet potentially meaningful to minority audiences."[19] What happened to Cantor during the late 1920s and early 1930s is similar to the management of the vaudeo persona that occurred twenty years later. Although early vaudeo stars were playing to largely urban audiences who would recognize ethnic tropes and allusions, performers such as Burns, Berle, Benny, and Caesar allowed tales of their origins and urban sensibilities to speak softly for their Jewish heritage, so as not to offend rural and suburban viewers.[20] Even though, as the Jewish historian Arthur Hertzberg reminds us, Jews themselves were becoming a part of suburban postwar life and entering into a period of rather intense assimilation.[21] The portrayal of stars' domestic lives was an essential extratextual element of the production of star personas during this period. It was particularly important for television stars, as the industry was trying to construct television as a family-friendly entity that rightfully belonged in the center of American home life. Although the popular press often initially portrayed the married vaudeo male as the head of his own happy family, this narrative often had to be rewritten in light of domestic discontent. Benny and Burns, both well known in the broadcasting industry from their success on radio, were easy to portray as family men because

Fig. 3.3 Milton Berle and his daughter Vicki on a November 21, 1950 cover of *Look* magazine.

both worked closely with their wives and adopted children during their broadcasting careers. Berle also adopted a daughter during his tenure as "Uncle Miltie," but his role as 1950s patriarch was tainted by marital problems.

In early 1950, *Redbook* reported that "the domestication of Milton Berle has proved a challenge even to the persuasive resources of his wife, a former showgirl billed as 'Joyce Matthews, the prettiest girl in America.' There have been troubled moments in the union, including a divorce."[22] After divorcing Berle for the second time, in 1952 Matthews attempted suicide in the apartment of her married lover, the theatrical producer Billy Rose.

In light of the threat such scandals posed to the stars' credibility as "happy family men," it was far more advantageous for publicity agents to emphasize the performers' extended families over their nuclear ones. Vaudeville was a handy tool in this process because it provided the appropriate backdrop for the typical American success story. Moreover, cultural memories of the massive influx of immigrants arriving in the United States in the early 1900s were potentially stimulated by the move of more than one million Americans from cities to suburbs during the postwar era. Douglas Gomery elucidates the connection between these two significant demographic shifts:

> To appreciate the scope of this internal migration compare it to the more famous transatlantic movement from Europe to the United States around the turn of the century. In 1907, when migration was at its peak, more than one million Europeans landed in the United States. This was precisely the magnitude of the suburban migration of the late 1940s and early 1950s.[23]

The press often described vaudeo performers as having successfully renegotiated their working-class positions through hard work, humor, and a kind of ethnic pluck. They were providers who enabled their often large families to escape the poverty of urban immigrant neighborhoods. Berle, Burns, and Caesar, all child vaudeville stars raised in poor Jewish neighborhoods, were quick to describe how their families had been integral to their success in the theater. They variously worked with their siblings, were pushed by their strong-willed mothers, or were inspired by the wish to provide for their large working-class clans. Many told stories of how they resisted the Old World traditions of their immigrant parents in order to better succeed in American society. Burns related his ambivalence toward the religious voracity of his cantor father, and Benny repeatedly told how his father hit him in the head with a prayer book as punishment for arriving late to Yom Kippur services.[24] Rubin argues that

> the contrast between unpleasant Jewish childhoods and high American aspirations characterized the life of most Jews growing up between the 1890s and 1930s . . . powerful emotional force pushed them toward success in American society and away from Jewish identity. But equally

ashamed of abandoning it, they often retained a strong sentimental attachment for that background.[25]

Berle, whose mother, Sandra, regularly appeared in his television act, had been discovered at the age of five when he won five dollars in a Bronx talent contest for mimicking Charlie Chaplin. His early years as a child actor in vaudeville and in silent film were often portrayed in the popular press as a struggle to keep his Harlem-based family together. Describing his father as ineffectual in supporting his family, Berle often told stories of his mother's undying support for his career and of how he helped his financially strapped family. Sandra was a constant presence in Berle's life and performances and acted as a continuing reminder to the public of his troubled beginnings and of his unending mother-love.

On television, Uncle Miltie was also cleverly crafted as a family entertainer, not only through his relationship with his mother (who was often in the studio audience) but also through his implied relationship with viewing families. He always had special goodnight greetings for the children in the audience and did television specials and benefits for children. In one such special, *Uncle Miltie's Christmas Party* (1950, NBC), water, powder, and pies were thrown in Berle's face. Although this was a rather common vaudeville joke, it also allowed Berle's masculinity to be subjugated, helping to make him everybody's favorite uncle: a nonsexual, nonthreatening member of the family.

Vaudeo stars also emphasized their personal relationships with one another, which often were described as close-knit kinships. Countless articles in the popular press told of dinner parties, card games, and golf rounds attended by vaudeo stars and their nuclear families. These events were not understood as business-related but as gatherings of close friends. Furthermore, they were described with the reverence and fervor usually reserved for blood ties.[26] Jack Benny was portrayed as the center of this off-hours social group, acting as host and introducing many of its key players. Danny Kaye, Fred Allen, and Cantor were all regular visitors to the Benny household, but it was Burns and Allen who were Benny and his wife's closest friends. After Benny's wife, Mary Livingston, introduced Allen to Burns, the foursome became virtually inseparable. Many of the articles that were supposed to be about one of the comic couples mentioned the other, describing Livingston and Allen's shopping trips or Burns and Benny's constant attempts to make each other laugh. Although it was said that Burns made Benny laugh the hardest of any of his comedian friends, Burns told *Cosmopolitan* that "because Benny is such a great audience, in fact that greatest audience, he has everybody in the business working for him. I know lots of comics would rather do an hour's entertainment for Jack, in Jack's living room, than play a week in a big theater in New York."[27] Even Benny's marriage connected him to another comic family, as Livingston (née Sadie Marks) was related to film comedian Zeppo Marx. Close bonds among vaudeo stars were privileged in these stars' promotional campaigns and served to reassert the stars' affiliations with extended familial networks.

References to vaudeo performers' extended families and their ethnic working-class upbringings resonated with the pasts of many urban television viewers. But such allusions may also have stimulated sentimental memories of extended familial structures (which were now threatened by encroaching suburbanization) as well as earlier ethnic politics and processes of assimilation occurring during the 1920s and 1930s. In the years immediately following World War II, the vaudeville aesthetic reminded white Americans of a time when ethnic politics appeared, in retrospect, to be comparatively uncomplicated. Revelations of the Holocaust and fears associated with the Cold War came to weigh heavily on the minds of many Americans in the years following the war. (In retrospect, the turn-of-the-century waves of immigrants, which brought significant numbers of European Jews to the United States, experienced a relatively smooth assimilation.) In addition, the State of Israel was proclaimed in May 1948 as a safe haven for Jews. In light of the discourses on the Holocaust, the television vaudevillian, largely understood as "ethnic" and most often Jewish, was a comforting sight to many Americans who had only recently learned of Nazi Germany's attempt to exterminate the Jewish people. Henry Popkin, who complained bitterly about the lack of Jewish characters or performers in film during the immediate postwar period, wrote in a 1952 edition of the magazine *Commentary*:

> Whatever may happen in the future, one fact is encouraging; in television, as in radio during its heyday, there prevails what might be called the New York idea, that Jewishness is not freakish or embarrassing and there might as well be Jewish comedians as any other kind. Hence one finds the sort of Jewish reference, whether comic or not, that stands in refreshing contrast to the rest of our antiseptically "Aryanized" popular culture: a Yiddish phrase spoken, a Jewish dialect or intonation, an identifiably Jewish ironic quality. All of this expresses no attitude, breaks

no lances against anti-Semitism; it only recognizes one fact of experience: that Jews do exist.[28]

Despite the fact that subtle references to ethnicity may have been comforting to urban and Jewish viewers of early television, Jewish performers
were in many ways suspicious figures in American culture during the late
1940s and early 1950s. The implementation of the broadcasting blacklist
after the publication in 1948 of *Red Channels: A Report on Communism in
Broadcasting* revealed that many Jewish performers were unfairly targeted
as communists because of their ethnic heritage. For some, the Rosenberg
trial in 1952 was the pinnacle of the rhetorical confluence of anti-Semitism and anticommunism. Some Jewish organizations and leaders tried to
distance themselves from the stain of communism by engaging in a bit of
red-baiting themselves.[29]

A rather direct coupling of the Rosenberg case with the commercial
image of Benny was even united through the infamous Jell-O box that
Julius Rosenberg supposedly used as a password in his covert transactions.
In some accounts, upon exchange of the Jell-O box, either the phrase
"Benny sent me" or "Benny from New York sent me" was passed verbally
between the two men.[30] Even without this direct reference to "Benny,"
most Americans would have automatically connected the gelatin product
with the comedian. As Marjorie Garber points out, "So closely did American audiences link Benny and Jell-O that in a poll taken in 1973 of
middle-aged listeners asked to name his radio sponsor, most answered
'Jell-O,' although in fact Benny was subsequently sponsored by Lucky
Strike cigarettes for almost twice as long as he had been by Jell-O."[31] Garber goes on to note that Benny's Jewishness was significant in the cultural
resonance of the Jell-O aspect of the Rosenberg trial, as was the product's
complicated relationship to Jewish law (i.e., the nonkosher origins of gelatin but its eventual kosher certification after manufacture).

Events such as the publication of *Red Channels* and the Rosenberg trial
belie the complicated, and frequently treacherous, signifiers that often
piggybacked on the codes of Jewishness. So, although Jewish ethnicity
may have been essential in the revitalization of the vaudeville aesthetic
and, subsequently, in the naturalization of the television medium as a
legitimate form of cultural production, it was difficult to acknowledge a
performer's Jewish ethnicity routinely or overtly. Instead, ethnicity was
commonly moderated through the codes of urbanity and immigrant child-
hoods and family.

Yet another significant way to reference obliquely Jewishness was through the performance of masculinity. The male vaudeo star's multivalent sexuality had the potential to activate long-standing assumptions about the nature of his ethnic background.

Mobilizing Ethnicity by Troubling Masculinity

Although his on-screen comedy was based largely on wordplay, insulting jokes, and topical gags, Berle's costume acts were the most popular segments of *Texaco Star Theartre*. The comedian was known for his female impersonations of, for example, Cleopatra, Carmen Miranda, and the opera star Dorothy Kirsten (who was so insulted by his portrayal of her that she took him to court). As the *New York Times* wrote in a 1990 retrospective article, Berle "was a man who wasn't afraid of a dress and for four years he owned Saturday night."[32] Although his cross-dressing was a long-standing practice in vaudeville and was not often directly described in the press as being sexually transgressive, a 1951 article in *Sponsor* implied that sponsors and broadcasters were aware of such tensions:

> Some eyebrows in the trade are up more or less all the time at TV's tolerance of "swish" routines and impersonations. They think that's going too far, projecting a special brand of big city tenderloin into the family circle. Other observers are more relaxed about the "swish" stuff, think it will be interpreted as nothing more than a spoofing of sex characteristics.[33]

Berle's propensity for donning a dress would have been especially potent when considered in relation to his ethnicity because many theorists have argued that femininity has long been considered a concomitant feature of male Jewishness. Citing examples from the early church to Nazi propaganda films, Garber notes that "not only sartorially, but also scientifically and theoretically, the idea of the Jewish man as 'effeminate' as well as 'degenerate' has a long and unlovely history in European culture."[34] The characteristics that seemed to set Jewish men (particularly those of the Hasidic sect) apart—the way they spoke, dressed, gestured—were regarded by many gentiles as not only foreign but "woman-like."[35]

Daniel Boyarin, in his detailed analysis of Jewish masculinity, suggests that the European conception of the Jewish man as half man, half woman was asserted by the Jewish culture itself. It is a reaction to the marginalization of the culture, as "Jewish society needed an image against which to define itself and produced the 'goy'—the hypermale—as its countertype, as a reverse of its social norm."[36] Boyarin goes on to suggest that this alternative gender typing is a historical strategy for cultural survival: instead of imitating their oppressors, Jewish men embraced the antithesis of Anglo "hypermasculinity."[37] Although Boyarin never specifically mentions any of the vaudevillians who worked on television, Garber ultimately finds Berle's form of cross-dressing a strategy that reasserted feminization of the Jewish man. She claims his drag impersonations were a "prerogative" of power because he chose to cross-dress and, in doing so, he directly confronted the stigmatization of the Jewish man as feminine. Garber writes that "for a Borscht Belt comedian like Milton Berle, whose routines so often included a drag act, to cross-dress for success, recuperating, however unconsciously, this 'feminization' of the Jewish man, and deploying gender parody is an empowering strategy."[38]

Beyond referencing the feminization of Jewish masculinity, another determinant of Berle's "feminization" was his relationship with his mother. His publicity material, which focused primarily on his childhood performances as a working-class Harlem native, repeatedly emphasized his overbearing yet loving Jewish mother. In an article entitled "My Son, Uncle Miltie," published in the *LA Examiner* on Mother's Day 1952, Sandra Berle writes, "As for a good many years, wherever he went I did go, getting the bookings, fighting the would-be managers and agents and talent developers, cooking over a can of Sterno in hotel rooms for the both of us, living out of a trunk. Those were the hard days, and then came the good ones, all these good ones."[39]

Berle's attachment to his mother managed to domesticate him, to connect him with family life, yet at the same time it demasculinized him further, even if his attachment was couched in the comforting discourse of immigrant culture. The brashness and sexual innuendoes of Berle's live performances were tamed, and the long-standing stereotype about Jewish mothers and their sons was clearly reinforced for the television audience. Berle told Gladys Hall, a writer for *Radio/TV Mirror*, that he always included Sandra in his life and performances because of all she did for him during his childhood. "Some forget, when they grow up what their mothers did for them and gave to them. I don't forget. I remember . . . all the sacrifices she made for me and the things she went without so that I could succeed. I don't think there is ever enough that you can do for your mother. Ever."[40] In a 1989 interview, Berle expressed concern over his feminized image. When asked why he had so many extramarital affairs, Berle responded by saying, "Maybe I had to prove my manhood to the

Fig. 3.4 A cover of Milton Berle's orchestral album "Songs My Mother Loved."

outside world that always saw me with my mother and wearing dresses in my act. Is she his 'beard'? Is he gay? Maybe that's why I played around so much."[41]

In his pairing with Dean Martin, Jewish comic Jerry Lewis was similarly demasculinized. Much of their comic chemistry as a team came from the way that Martin's lothario persona would play off Lewis's ambiguous man-child character, creating what Andrew Sarris has described as their "persistent sexual hostility."[42] In part, Lewis's demasculinization was because of his gangly, twitchy body, elastic face, and odd vocal utterances, which moved him out of the realm of heterosexual romantic hero and into a more ambiguous gendering. But it also was his ethnicity as it was othered against Martin's Italian machismo that put those characteristics in stark relief. Lewis has said that he borrowed much of his performance style from his father, Danny, who had performed in vaudeville and in the Borscht Belt, and this is certainly visible in his television performances.[43] He was performing a modified version of type of historically Jewish style

Fig. 3.5 Martin and Lewis on *Colgate Comedy Hour.* Library of American Broadcasting.

of comedy as he engaged with and responded to (often ad-libbing) Martin's smooth, straight man, Italian crooner. A number of scholars have remarked that Martin and Lewis seemed to resemble a couple at times, with all the homoerotic implications placed onto Lewis. Ed Sikov describes him as a "jester in a court of sexual panic" and believed that during

Fig. 3.6   Martin and Lewis for NBC radio. Library of American Broadcasting.

the postwar era, "Lewis could express the (for want of a better word) homo-eroticism that could no longer be denied on the screen but could scarcely be stated forthrighty in this era of officially-sanctioned gay-bashing."[44]

Jack Benny, whose vanity, posture, and gestures could have been con-strued as feminine, was a tamer version of the Jewish male comic.[45] Unlike Berle, Benny often avoided the use of the puns and slapstick moves of traditional vaudeville routines, choosing instead to place his radio and tele-vision narratives within a situational context. However, Benny's use of the cheapskate stereotype, his violin playing, and his reliance on self-disparaging humor were definite components of both his vaudeville performances and his ethnic background. According to Hilmes and Margaret McFadden, Benny's feminization originated in his radio program. Hilmes writes:

> Jack comically violated all the norms of American masculinity. Obviously wealthy but unable to spend money, thinking himself the pinnacle of masculine attractiveness but unable to interest women, suave and debonair but unable to handle simple situations, the authoritative host of a major radio program but unable to command the respect of his employees (and, later, a white man totally dependent on his black servant, in a relationship with strangely homoerotic implications).[46]

McFadden argues that the construction of Jack and Rochester as an interracial couple is undermined by both characters' pursuance of women within the text of the TV program.[47] She believes that the subversive potential of their relationship is contained, as "their aggressive heterosexuality, like that of the other men in the cast, obscures their all-male family and thus makes it acceptable. Further, this ambiguity allows the construction of a family where certain non-threatening women and some feminine qualities in men are permitted, but where no culturally idealized wives and mothers intrude."[48] Denise Mann contends, however, that Benny's "aggressive heterosexuality" in relation to the female film stars that appeared on his show actually feminized him further, at least momentarily, by placing him in the role of ogling fan.[49]

Still, Benny had little trouble fitting within mainstream America's conception of a domestic entertainer. This was, in part, the result of a consistent publicity campaign that depicted him as a benevolent family man. Benny's stage persona as a curmudgeonly, vain penny pincher was disavowed in numerous press releases and publications about his real life during the late 1940s and early 1950s. Unlike the publicity surrounding other vaudeo stars of the period that worked to conflate a performer's on-screen personality with that of his off-screen identity, articles and press releases about Benny often denied any direct connection between his two identities. This may be because the situational narrative of *The Jack Benny Program* required Benny, unlike Berle and Caesar, to remain in one character during the entire broadcast. And, although the Benny character could be endearing at times, he also could be quite despicable. To prevent broadcast audiences from believing that Benny was really as vain and cheap as his on-screen character, Benny and his publicists carefully

Fig. 3.7 Jack Benny. Author's personal collection.

managed to retain the most positive aspects of Benny's broadcast persona while denying the "reality" of the most offensive aspects of that image. Stories about his rise to fame as a child violinist, as well as about his generosity and his role as a dedicated family man and loyal friend, were circulated as means to this end. In a 1944 article in *American Magazine*, Benny's radio personality was suggested to be a great psychic burden to the "real" Benny. Jerome Beatty writes: "If you or I made $10,000 every Sunday we wouldn't care if the world thought we were Jo-Jo, the dog-faced boy. But Jack Benny, at 50, takes life and public opinion as seriously as a candidate for the U.S. Senate and spends most of his spare time chasing his shadow, trying to stamp it out."[50]

In a 1949 CBS press release, Mary Livingston claimed, "Jack has his own hair and teeth, is not anemic, is in perfect physical condition, can play a pretty good violin and in my opinion, and in the opinion of almost everyone who really knows him, is the greatest guy in the world."[51] In addition, Benny often played with the constructed nature of his broadcast character and public image. For example, a 1951 layout of Benny and his family in *Look* parodied the supposed "real-life" publicity photos that Hollywood stars typically had taken of them in staged domestic settings. Stanley Gordon writes facetiously in the introduction that "from now on, Benny intends to be photographed like the Hollywood stars in movie magazines—in cozy family scenes to warm the hearts of his many fans, pictures to show that the Bennys are, after all, 'just plain folks.'"[52] Jack, daughter Joan, and Mary are in the accompanying photos of them doing the dishes, working on Joan's homework, cleaning the car, singing at the piano. A caption for a shot of Jack hanging laundry on the line reads, "The domestic type: Jack hangs out a few things in the laundry yard (A full-time laundress is really the one who does the family wash daily.)." Another caption, accompanying an image of Jack, Mary, and Joan sitting around the piano singing and playing the violin, reads, "After dinner, the Bennys always gather round the piano for a musicale. (In fact, Jack keeps violin in bathroom, plays only on his CBS program.)."[53]

By pointing out the media fabrication of stars and their home lives through these joking asides, Benny destabilized the authenticity of his fictional persona. In so doing, he was able to pick and choose which characteristics he wished to emphasize and which he would deemphasize. He could remain the vain fall guy on screen or in radio, but he was also able to claim that he was still "authentic" in the way that the public would value (i.e., he was trustworthy, caring, and generous), thereby making him an acceptable role model and product pitchman.

Although the assumptions of Jewishness and a peculiarly feminized masculinity mobilized around Benny and continued to make him popular throughout his career, they were also denied, perhaps in an effort to bolster his fiscal and social standing in the 1950s. By recreating an image of the hard-working, generous, and often guilt-ridden Benny, the home viewing audience could be more comfortable with his extraordinarily glamorous lifestyle—and, specifically, with his much-publicized post-talent raid salary. But even the revised Benny, although now in recoil from

Fig. 3.8   After years of being 39, Jack Benny Turns 40. *Collier's* February 19, 1954.

some of the more long-standing Jewish stereotypes of his earlier persona, was still entrenched in the mythic struggle of a working-class boy made good. This struggle, as I have shown, was firmly imbricated with nostalgia for the immigrant and extended family and for domestication of the vaudeville aesthetic.

The form of ethnic (read Jewish) masculinity that helped shape the careers and personas of early television comedians had a limited utility in the subsequent development of the medium's industrial and cultural operations. For a brief period, coinciding with the years of the Federal Communications Commission's freeze on licenses, performances by men who cultivated a flexible yet historically defined and culturally determined gender identity dominated television. The industry was able both to exploit and to counter the signs of an alternative form of masculinity during its campaign to enter the American home by mobilizing the cultural memory of an audience unsettled by recent social upheavals and the breakup of the extended family.

By the mid-1950s, the variety show was clearly in decline. As many television historians have noted, this was, at least in part, because of the lifting of the FCC's freeze on licenses and the changing demographics of the television audience. No longer primarily urban and as ethnically mixed, the audience was now increasingly middle class and lived in rural as well as suburban areas. The cultural references made by Jewish vaudeville comedians ceased to be familiar to a large part of the audience.

As I will discuss in further detail in the following chapter, after 1951 the television industry also was forced to appease audiences and critics who were becoming bored with vaudeo performers and the sponsors who tired of paying their exorbitant salaries. Consequently, the frequent and forceful cries of sponsors and trade critics for new faces and formats had a significant impact on the industry's eventual embrace of the sitcom. More concerned with the repetitious nature of vaudeo than with its inability to resonate with viewers outside the East Coast, some sponsors and critics expressed concern as early as 1950 about the long-term viability of the vaudeville performance style and the age of its television headliners. Sponsors were also deeply worried about the financial implications of what *Variety* dubbed the "Talent Crisis," a period in the early 1950s when the industry, feeling the strain of its prior disregard for talent development, allowed vaudeo star salaries to skyrocket and dramatically increase the overall production costs of the variety format. As a result, the sitcom, with its emphasis on narrative and its reliance on a regular cast (rather than expensive, high-profile guest stars), looked like an economically viable venture for sponsors and independent producers alike, especially since successful examples were already on the air.

Although traces of the vaudeville tradition were still present in television comedy, plot-driven, "realistic," suburban sitcoms dominated network program schedules by the late 1950s and early 1960s. This format offered viewers a new vision of domesticity, identity, and consumer culture that differed dramatically from that of early variety shows and ethnic sitcoms. George Lipsitz suggests that "for Americans to accept the new world of 1950s consumerism, they had to make a break with the past."[54] In the case of television, the industry and its audience were ready to break with vaudeo, as the format's references to prior traditions of ethnic masculinity and immigrant life were no longer culturally relevant to viewers now eager to embrace the "good life" promised in suburbia.

CHAPTER 4
"TV IS A KILLER!"
The Collapse of the Vaudeo Star
and Television's Talent Crisis

By the early 1950s, the broadcasting industry had begun the process of overhauling many of its production and business practices. Radio was fast becoming more about the musical tastes of local markets and less about the national presentation of narrative. In 1952, the FCC lifted its four-year freeze on licenses, and as a result new television stations proliferated in areas outside of the Northeast, significantly altering the demographic makeup of the television audience. During that same year, the television industry established its censorship manual, the Television Code; NBC and CBS invested heavily in the construction of studios in Los Angeles; telefilm became a viable option; coaxial cable enabled coast-to-coast simultaneous transmission; the cost of television production was rising to unforeseen heights; and the television, film, and radio unions engaged in a very public battle over the representation of television actors.

One result of these tumultuous years of transition in the early 1950s was that the vaudeo star and variety format began to decline in popularity. As mentioned in Chapter 2, critics were expressing concern about the long-term viability of the vaudeo star as early as 1949. "Television programming is beginning to look like a Palace bill of 20 years ago. What are we going to do when these name stars pass their peak or feel it's time to retire?" asked *Variety* in an article reporting on what it had dubbed "the new talent crisis."[1] The vaudeo performers' age as well what some perceived as stale material and comedic styles became issues early on as commentators recognized the limited utility of the vaudeville aesthetic. Industry surveys and studies seemed to show that audiences, too, were tiring of the formats and faces that dominated network television during its first few years. In 1949, *Variety* claimed that vaudeo stars were already beginning to bore audiences: "Bearing out the old maxim that familiarity breeds contempt, the first qualitative study yet made on television viewer reaction to TV performers and programs has confirmed the belief that a too-familiar face, no matter how entertaining at first, may become irritating through constant viewing."[2] The paper also argued that television could no longer rely on its unique qualities to attract audiences because "[Video] has grown in such gargantuan proportions that the novelty appeal has already disappeared for hundreds of thousands of lookers. Just seeing objects animate in one's living room no longer is unusual. The values have asserted themselves."[3] Hollywood producer Hal Roach blamed a dearth of industrial initiative for television's failure to produce new stars. He argued that television could not keep poaching its talent to fill its 280 hours a week of programming. "If Hollywood won't give you TV broadcasters its stars, you'll have to go out and find stars for yourself. After all, where do you think all of the Hollywood stars originally came from?" asked Roach. "They weren't born in the studios."[4]

In the early years of the 1950s it was obvious to most that the television industry needed to revamp many of its early narrative and economic assumptions. Although vaudeo was intrinsic to television's initial dissemination, it became increasingly clear that it could only temporarily sate television's voracious appetite for programming. Moreover, as talent costs skyrocketed and audience demographics began to shift, sponsors and viewers began to lose interest in the vaudeo format altogether. One immediate answer to the problem was to increase production of amateur hours, as they were cheap to produce and provided an opportunity to try out new television talent. CBS executives, however, interested in a long-term solution, set their sights on the suburban sitcom, believing the format's emphasis on narrative and reliance on a regular cast (rather than expensive, high-profile guest stars) would standardize costs and attract a lucrative and loyal viewership.

The broadcast historians who have studied the larger shifts in production and narrative strategies of the early to mid-1950s have not considered the impact that stardom had in the industry's move from live, single-sponsored variety programs to telefilmed, independently produced, suburban sitcoms. In the following pages, we'll see that the issues central to the talent crisis were seminal agents of the variety-to-sitcom transition and that they were essential factors in increased network control of programming and the eventual evisceration of the single sponsor.

The Collapse of the Vaudeo Star

Beginning in 1951, the popular and industry press alike began to report on the toll that television work exacted from variety stars. Most top-name variety talent were under such stress from "TV's exacting demands" in terms of scheduling and live performance that they ended up in the hospital with nervous exhaustion and physical distress (including ulcers and heart attacks).[5] In 1952 *Variety* compiled a "TV Casualty List" that included such stars as Milton Berle, Jack Benny, George Jessel, Red Skelton, Eddie Cantor, Fred Allen, Red Buttons, Ed Sullivan, Dean Martin, Jerry Lewis, and Jackie Gleason, all of whom suffered such maladies.[6] "When I collapsed from the strain and overwork of being jolly for almost seven years, I found myself in excellent company," wrote Berle in an article entitled "TV Is a Killer!" He continued, "Hospital beds nowadays swarm with human wrecks from the TV wars—comics who sign up for 13 weeks with options, only to get a 24-hour virus with ulcers."[7] Max Liebman, a writer for *Your Show of Shows,* expressed an even harsher attitude toward the work that the medium required: "I am dead certain that TV is the toughest, back-breakingest, ulcer-breedingest entertainment medium in existence—a fascinating monster that devours material, tortures talent, sears souls and paralyzes the participant."[8]

Live television did take quite a bit out of its talent—writers, producers, directors, and performers alike. Vaudeo required at least an hour of new scripted and memorized material each week. Although radio performers were prepared for the weekly production cycle of television, they had little experience in memorizing lines and being conscious of their physical movements for the camera. Ex-vaudevillians were used to physical displays of humor and performing night after night, but on the vaudeville circuit they would repeat their same acts to new audiences. In television, production staffs were aware that once an act or joke had been used one week, it could not be used again—at least for a while.[9] Cantor, who suffered a heart attack in 1952 during his run on television, recognized two years earlier that television would bring him a significant amount of stress. He went into the medium because it offered him the same excitement and anxieties that he had experienced on stage during the opening nights of his "Ziegfeld Follies." Yet Cantor noted that television demanded more from him than stagework did: "Once we opened in the 'Follies' we had nothing to worry about for a year. Now we do a show and start right in worrying all over again about the next one. If that doesn't take something out of you, nothing does."[10] Edgar Bergen, who had witnessed the collapse of many of his friends and colleagues, admitted that he was afraid to work on television because of the way video was "destroying talent." Bergen believed that "Networks and agencies have done practically nothing to protect comics from becoming tele casualties and hurting their value in radio."[11]

One answer to the dilemma posed by television's intense demands on its labor was to cut the number of on-air hours for each program. Starting in late 1951, comics pushed for alterations to their programming schedules. Most of them pressured the networks into allowing them to perform for a half hour every other week. In fact, *Variety* reported that the movement for this type of scheduling was particularly tenacious, "For once [comics] are in complete accord . . . they're practically ganging up on the networks to force a showdown for the [1952–1953] season."[12] Comedians who were exhausted by weekly hour schedules were joined by performers such as Danny Thomas and Jimmy Durante, who were each contracted to do only one hour-long program every month. Even though Pat Weaver thought he had headed off such a problem by rotating comics and contracting stars to appear no more than every other week on the networks, comics still complained. They believed that an hour-long show was simply too taxing on a single performer and demanded that NBC allow them to each do a half hour show every other week.[13]

Television stars found that television work also prevented them from working in other media. Berle, Godfrey, and Phil Silvers complained loudly and publicly about the stress they experienced from trying to work in television, radio, and stage simultaneously and eventually all chose to stick exclusively to television.[14] Unlike their years in radio, when a star could moonlight rather easily in films or on stage, television required complete dedication and commitment. On realizing this, top name talent would often demand more money from the networks in order to make up the pay they were losing by not participating in other entertainment media. Martin and Lewis as well as Donald O'Connor were forced to cancel their turns as hosts of *Colgate Comedy Hour* during the 1952–1953 season on the advice of their doctors. They found they could not do television and film

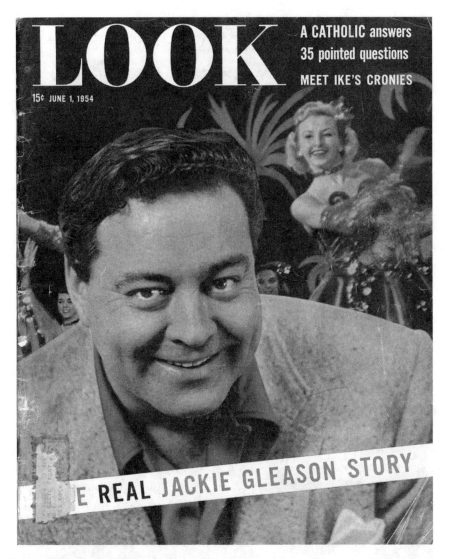

Fig. 4.1 Jackie Gleason on the cover of *Look*. June 1, 1954.

work at the same time and, unlike other television stars such as Berle, chose to focus solely on their careers in Hollywood.

Gleason and Caesar seemed to be prime examples of the dire consequences of television variety's arduous work as the press intensely focused on their resulting physical and psychological vulnerabilities. During their initial years on television, both performers had been perceived as affable, skillful comedians enmeshed in the working-class nostalgia of early television comedy. They were strong men with large personalities, but as

Fig. 4.2 Jackie Gleason. Library of American Broadcasting.

they became more well known to audiences, press articles on the pathos that plagued their comedy began to appear in rather large numbers. Gleason was most often described as excessive in his eating, drinking, and womanizing.[15] His perceived inability to control these aspects of his life were seen as consequences of both his professional perfectionism and the grinding drive of television work. These dysfunctions and excesses, in other words, were renegotiated into an extension of the vaudevillian rags-to-riches story. Television, it would seem from these press accounts, in reviving vaudeville had also revived the immigrant work ethic. Now, however, this carried serious ramifications for the physical and psychological welfare of its labor. Although it was well known that even Berle and Benny were hospitalized during this time for exhaustion, it was Gleason and Caesar who were seen to have taken the worst of the psychological toll. *Life* profiled Gleason in 1955: "Every comedian has found the wear and tear of producing a weekly TV show almost more than the human physique can stand. . . . Gleason's rehearsals are even more agonized than most as he seeks unceasingly to add the last speck of polish."[16] His peculiar combination of perfectionism and excess was, it seemed, the primary source of his psychological unrest. Gleason stated publicly that he never wanted to enter into psychoanalysis to get at his problems, for reasons which were obliquely attributed to his strict adherence to Catholicism and his stoic working-class ethics. "I didn't need a psychiatrist to tell me that one of the things that gives me insatiable hunger and causes me to pile on blubber is to be under constant stress and strain," Gleason told an interviewer, "Whenever there is strain or insecurity involved—and television manufactures those things incessantly—you find yourself turning to the icebox the way a breast-fed baby turns to his mama."[17]

Unlike Gleason, Caesar actively promoted the use of mental health care. Perhaps envisioned as a part of his Jewish cultural heritage, Caesar's foray into psychoanalysis was widely touted as a sound deterrent to the dysfunctions inherent to a television career. In his *Look* article, "What Psychoanalysis Did for Me," Caesar discussed the difficulties he had as a result of the success of his NBC *Your Show of Shows*:

> On stage, I could hide behind the characters and inanimate objects I created. Off stage, with my real personality bared for all to see, I was a mess. It was difficult for me to establish a normal, healthy relationship with anyone. I couldn't believe that anyone could like me for myself. I thought people were around me because they wanted something from me.[18]

Although Caesar told the press that psychoanalysis had helped him resolve his difficulties with the disjuncture between his public and private identities, he and Gleason both continued to struggle with such issues throughout the 1950s. According to reports, they seemed to be victims of a medium that took a traditional form of entertainment (vaudeville) and accelerated its pace to an unhealthy level. The rhetoric of the American work ethic that was part of immigrant nostalgia appeared to turn in on itself at this point in the descriptions of the backstage lives of performers such as Caesar and Gleason. The vaudeo stars' intense labor was no longer the reason for their success; rather, it was the reason for their demise. It also was apparent in the mid-1950s that the age of the vaudeo star had become a factor in both their deteriorating health and in the decline of the format itself.

The networks believed that they could ease some of the tension between them and their talent by building studios in Los Angeles. In doing so, they could provide talent with a more hospitable climate, larger studio space, and close proximity to Hollywood. CBS and NBC both began construction of immense studios in Hollywood in 1951 with an eye towards interpenetration between broadcasting and film industries. Both networks were cognizant of the impact these studios would have in attracting and retaining stars.[19] In his 1992 autobiography, former NBC president Pat Weaver reflected that a driving reason behind the move to Los Angeles was the talent: "We weren't likely to attract the stars we wanted if they would have to fly to New York to work for us, especially if by that time they could stay at home and work for CBS."[20] George Rosen pointed out that moving top network comics out West would take care of a number of talent issues with which the industry was struggling in the early 1950s:

> Shift of the comics to Hollywood will ease network tension in relation to the studio space situation, for the rehearsal time required in these full-hour extravaganzas has created a serious bottleneck and shunting of other shows to empty lofts and even remote areas of New York. Thus, the TV situation will parallel that of radio some years back, when the lure of films, climate, and general living conditions gave Hollywood No. 1 ranking on top AM-originating programs.[21]

Yet a few comics chose to stay in New York.[22] As a result, NBC's comic lineup was split (albeit unevenly) between the two coasts as Cantor, Hope, Martin and Lewis, Durante, Thomas, Jack Carson, Ed Wynn, and Skelton moved their broadcasting productions to Los Angeles, while Berle and Caesar remained in New York.[23]

Although live television was still the industry's preferred format for comedy shows during this time, the advent of telefilm would eventually provide a respite for television talent. During the years in which vaudeo stars were collapsing from exhaustion, the broadcast industry was recognizing the many ways in which its dependency on top-name variety talent and the live format was putting pressure on the industrial structures that had carried over from radio. It was becoming obvious that the system was no longer working as talent complained about conditions, sponsors and ad agencies balked at rising production costs, networks garnered more control over production and talent, and audiences and critics called for new formats and faces. Networks promised changes to temper the unrest, but inevitably they would have to admit that the system needed a major overhaul. One of the first areas in which all the major issues at stake in the industry's transitional years came to a head was around the rising cost of talent and the financial and creative power that many top stars were wielding in the industry.

High Talent Costs and Star Censorship

The trade press during the early 1950s was replete with articles decrying the high cost of star salaries. At the time when CBS and NBC were battling for the top stars in television and had signed Hope, Berle, and Benny to almost lifelong contracts, sponsors fretted about the exorbitant budgets required to secure top-name talent, as well as the control that networks were exerting through their relationships with such performers. In a speech to the radio-television production session of the 1951 American Association of Advertising Agencies, TV director of the William Morris Agency, Wallace S. Jordan, stated, "Stars welcome the long-term network contracts for the security they provide, while agencies and sponsors don't sign the talent for such long-term deals."[24] Pointing out that the control of time and programs on video was interlocked with the control of talent, Jordan acknowledged "the apparent unwillingness of agencies and bankrollers to take a chance on new and untried tele talent" and suggested that sponsors and agencies should earmark a large part of their TV budgets for developing relatively new performers. Although some sponsors have tried new faces, Jordan said, they generally don't give them enough support, and the result is often a "13-week turkey."[25] Rosen also criticized the networks for investing so much into so few. He argued that the networks would be well advised to put a larger percentage of their money toward program development:

The fancy-priced jockeying between the two major television networks over the past couple of weeks for top personalities, reminiscent of the "golden era" in radio, is causing widespread alarm within the industry. It's the prevailing feeling that the networks, notably NBC and CBS, by perpetuating an evil that is sending talent costs spiraling to new highs, have learned little from the hues and cries when radio was in full bloom; instead of investing coin in experimenting with new program ideas and formulas to keep a medium alive, they are channeling their energies and bankrolls toward a handful of personalities.[26]

Talent agents and stars defended themselves by arguing that the rise in salaries was an inherent consequence of the requisite demands of television work. "A Morris agency exec says that high talent costs are a result of the performers not having time to work in any other entertainment medium and caused by competition between shows," reported *Variety* in 1951.[27] Live television tended to require more out of its talent than any other medium. Performers were constantly reminding networks of this and threatening to quit if they were not adequately rewarded for their efforts.[28]

Sponsors had been cautiously eyeing the rising rates of television performers from the start of vaudeo's development. *Texaco Star Theatre* was an early leader in the salary hikes game. *Advertising Age* estimated that the program's talent-production cost for 1948 averaged about $9,500 a week and that by September 1950, the cost was up to $35,500. (In comparison, the 1950 weekly talent-production costs—not including time charges—totaled $15,000 for *Godfrey and Friends* and only $7,500 for *The Goldbergs*.[29]) Adding the $20,000 charge for network airtime brought Texaco's total cost to approximately $65,000 a week.[30] In March 1951, NBC penned a thirty-year contract with Berle wherein he would work for twenty of those years and be paid a stipend of $50,000 a year for ten years after he retired.[31] A 1953 NBC accounting memo to David Sarnoff revealed that the network had committed $6.2 million, or 52 percent of its talent budget to six performers—Berle, Durante, Phil Harris, Dinah Shore, Caesar, and Bob Hope.[32] J.H.S. Ellis, president of Kudner, the advertising agency producing *Texaco,* was the first to decry publicly the rising cost of television production, although he was no longer paying Berle his salary. Other agency executives soon joined in, claiming that if television continued on this path it would "price itself right out of existence."[33] In April 1952, Colgate and Procter & Gamble, two of the biggest advertisers on television, publicly questioned whether their multimillion-dollar television budgets (Colgate was spending $6 million annually for its *Comedy Hour* on NBC) were actually paying off for them.[34]

Responding to agency and sponsor concern over rising production costs in the summer of 1951, the fourth-ranked Dumont network tried to seduce Texaco away from NBC with an offer to absorb up to $750,000 of the program's production costs. This maneuver was the first of its kind and hinted at the upcoming shift from sponsor to network control and financing of production. In the end, NBC was forced to match Dumont's bid and, in addition, the network guaranteed Texaco a "long-range exclusive franchise on Tuesday night at 8 period," which was considered "extremely important because of the growing scarcity of Class A time availabilities in TV."[35]

In 1951 high-profile guest stars on variety programs were averaging between $2,000 and $5,000 an episode, and national television stars were costing as much as $200,000 a year.[36] By the following year, the American Federation of Television and Radio Artists garnered a 12.5 percent rate hike in minimum pay scales for actors contracted to any of the four networks.[37] Agents, unions, and performers were acting on television's reliance on stars to attract audiences, and they knew that competition among the networks could lead to perpetual raises. Another reason for the rise in guest-appearance rates was the general acceptance of the medium by the Hollywood community. *Variety* reported late in 1949: "Video is no longer being shunned by the film colony on the double score of being an 'amateur' medium and a box-office threat." But, it added that although "Major studios have not about-faced to the point where they will permit valuable film star properties to do video shows," top freelance film talent rushed into television production.[38] Lured by relatively high fees during the immediate post-Paramount decree years when film work was less lucrative and plentiful than it had been in years past, some film stars made small forays into television, first with guest spots and later with regular roles. Christine Becker's research identifies the 1953–1954 season as the real turning point for top Hollywood stars appearing on television.[39] By then the relationship between the broadcast and film industries was becoming more synergistic; studios had begun to allow contracted players to appear on television; the medium was gaining more and more respect; and former film actors like Lucille Ball and Robert Montgomery proved that one could actually revitalize a career through television. "What does profit a cinema star to go into television?" asked *Time* in 1953. "TV pay has finally reached movie levels, and its multimillion audience is an attraction in a time of waning movie attendance. Best of all, it offers jobs during the dog days of Hollywood employment."[40] Joan Crawford claimed to be swayed by the quality of telefilm and said, "I find [television] extremely attractive, because it pays for itself and then becomes an annuity for my children. How else can you save money these days?"[41] Still, some film stars harbored reservations about the medium. Humphrey Bogart, who would eventually begin to make guest appearances in the mid-1950s, said, "I got a helluva good racket of my own. . . . I don't have the time and I don't trust the medium yet. . . . You watch that stuff some time. . . . Instead of being five foot eleven, you're four foot three. I'll wait until they get straightened out."[42]

The 1951 merger of ABC and United Paramount Theatres also promised to bring more Hollywood talent to television. In an interview with *Sponsor*, Leonard H. Goldenson, president of ABC-Paramount, promised that the merger would strengthen ABC's talent lineup. He pointed to the appointment of Robert M. Weitman, former director of Manhattan's Paramount Theatre, to the position of vice president in charge of program and talent development as a major factor in the predicted influx of major stars to the network.[43] As a result of such growing interpenetration between film and television, by 1954, when spectaculars were on the rise, talent and production costs for a one-time program would be as high as $250,000 when it carried names such as Judy Garland, Danny Kaye, or Bing Crosby.[44]

Although the rising cost of talent benefited performers, it was detrimental to sponsors in regards to their control over television production. The cost-benefit ratio of producing a television program had risen to a point that many significant sponsors were forced out of the medium. In fact, some believed that these sponsors should return to radio, as prices for advertising in that medium had begun to plummet. In May 1952, Rosen proclaimed that, "a new thinking is beginning to take hold in the advertising agencies. They're now telling their clients to 'stick with radio.' There are mounting fears that television is becoming a 'lost industry'—that unless something drastic is done, and done soon, the TV medium will price itself out of the advertising dollar field."[45] Yet radio was undergoing its own fallout from sponsors as it dropped many formerly successful network radio stars in order to produce cheaper programs such as quiz and give-away programs or to move into local advertising. Rosen continued, "The network radio jitters are becoming intensified. The outlook for next fall, executives concede, was never more bleak. The anticipated cancellation of major personalities has borne fruit, leaving wide-open gaps in cream time segments."[46] Because of pricing concerns as well as a significant fall in ratings, Chesterfield dropped Crosby, Heinz dropped Ozzie and Harriet Nelson, Philip Morris dropped Cantor, and Pet Milk dropped *Fibber McGee and Molly* in order to pick up Ralph Edwards's *Truth or Consequences* (saving them $3,000 per week).[47] This led to a general consensus among advertising agencies and sponsors that the financial and creative control of most popular prime-time network programming in both radio and television was gradually becoming the domain of the networks themselves. As a result, they were anxious to stop pouring money into such programs and to begin to find alternatives. *Variety* had predicted this move as early as mid-1949: "Most agencies with radio experience figure it's time to let the networks package the high-budgeted shows with the top-name stars. Then, if the stars do take over eventually, they will gain control only of the shows built by the webs. Such agencies, for their part, plan to concentrate on programs without name talent."[48] By the end of 1952, networks, agencies, and sponsors had accepted the inevitability of this trend. Yet, sponsors had unrealistic expectations for the type of ratings no-name talent could bring them. "While sponsors are calling for new comics, they're loath to hazard bankrolling an untried talent, due to big nuts their shows carry [*sic*]. They want 'ready made' ratings which only a name can bring."[49]

Another issue that piggybacked onto the struggle among networks, sponsors, agencies, and talent over wages was the amount of control that stars could exert over program content. Many stars were acting as producers of their own shows. Although they had to answer to agencies, sponsors, and networks, they were allowed a fair amount of authority over their program content. This was problematic for the sponsor, in particular, who wished to avoid controversy and to provide the most hospitable environment possible for the product. The high salaries that most of these stars received fueled the conflict between talent and the powers that drove the economic functions of the medium. Stars viewed their salaries as a confirmation of their centrality to the project of broadcasting while sponsors, agencies, and networks believed that those same high wages purchased the star and his/her autonomy.

Godfrey, who was acutely aware of the amount of money that he brought in to CBS, proved to be particularly troublesome in the area of content control. In one instance in 1950 Godfrey swore on the air. Consequently, two Midwestern affiliates refused to carry *Arthur Godfrey and His Friends*, and critics and politicians used Godfrey as an example of the moral debauchery of television. CBS tried to appease critics, affiliates, and the FCC by promising that this would not happen again, but it also had to admit that controlling Godfrey was extremely difficult, especially since so much of his program was ad-libbed.[50] Soon after the Godfrey debacle, FCC Chairman Wayne Coy stated in a speech at the Oklahoma State Radio Conference that, "When a comedian gets so big that his network can no longer handle him, then we have a case of the tail wagging the dog."[51]

Lynn Spigel argues that during the early 1950s, when government officials were focusing much attention on television content, vaudeo bore much of the brunt of the censorship debates.[52] Concerned about the bawdy antics, ethnic jokes, and sexual asides of vaudeo comedians, these politicians pointed to the format's New York sensibility as the root source of its offensiveness.[53] In *Television Program Production*, Carroll O'Meara contended that,

> What many entertainers fail to realize, actually, is that the areas containing the bistros, night spots, and bright lights are only a segment of America. And yet, somehow, they insist on broadcasting to the entire nation comic and other material which is definitely not acceptable in the average American home. . . . Our nation consists of 160 million citizens, most of whom live in small towns, go to church on Sunday, attempt to bring up their children decently, and do not regard burlesque as the ultimate in theatre.[54]

The public calls for censorship by public officials and cultural critics made sponsors nervous. Out of fear of losing their consumer base, they were forced to reconsider many of the basic tenets of vaudeo humor. A 1951 article in *Sponsor* claimed that television carried with it a greater risk of offending viewers than radio or film and recommended that sponsors censor problematic program content:

> Advertisers are by now pretty well briefed, or they ought to be, as to the everyday hazards lurking in racial jokes, dialects, characterizations, and superiority-inferiority situations. The pictorial factor in TV increases the danger. . . . So long as the sponsor's goal is universal good will for his products and services he cannot indulge in heavy-handed kidding and race-trait burlesque and then be surprised if Italians or Mexicans, or Irish, or Jews pass him by at the retail firing line.[55]

The ethnic and brash urban personalities that made vaudeo such a success in the late 1940s were now considered inappropriate for the increasingly national, middle-class audience. As a result, critics and politicians singled out the vaudeo star as an example of the tastelessness of network broadcasting. Many vaudeo producers responded by altering the structure and sensibility of their programs by introducing sitcom plots, erasing ethnic references, and making their programs more family-friendly overall. However, these changes would not ultimately save the vaudeo comedians who built early television. By 1953, Dick Powell, president of the Television Writers of America, was claiming that the "death of video comedy stars is being caused by censorship. . . . If Will Rogers were alive today, he would probably go back to rope-twirling."[56]

Beyond issues of sexual and ethnic content, sponsors also tended to shy away from anything that smacked of political controversy. This made them especially vulnerable to organizations such as the one that published the anticommunist publication *Red Channels,* which listed writers, producers, and actors in broadcasting accused of being communists and asked consumers to initiate letter-writing campaigns to have these individuals fired. This campaign was successful in pushing a number of actors off the air, including Philip Loeb and Jean Muir, who were fired from their programs as a result of their names being published in *Red Channels.* This made the relationship among sponsors, networks, and talent all the more tense. It also made the search for new talent more difficult. In an article, "How to Keep Reds off the Air—Sanely, "*Sponsor* complained, "Now, at a time when many believe [radio] must be more competitive, more experimental, more gutsy, than ever before, it is being asked to quietly accept the authority of a censorious blacklister. TV, just emerging as a major medium, is also asked to stifle itself just when it needs new talent."[57]

Still, concerns about political leanings did not prevent stars from exerting control over content even when it came to a program's commercial messages. Cantor, who had long been associated with a number of charitable organizations, agreed to promote Paper-Mate pens on his Colgate-sponsored program in exchange for Paper-Mate's contribution to one of his charities. After approaching Colgate with the idea and being rebuffed, Cantor took it on himself to mention Paper-Mate's name during his live program. However, NBC was able to mute Cantor's promotion as it was about to be transmitted. Another example of this is when Crosby angered General Electric. He went against the sponsor's wishes and featured a "strip-tease tassel tosser" on his family program.[58] These incidents led *Advertising Age* in 1954 to call for producers and sponsors to usurp the defiant power of television talent, using extremely blunt language to do so:

> Probably the greatest lesson from these experiences is to remember that talent must be thought of not as "people," but as chattels with prices on their foreheads, like any other machine working for a manufacturer. They must be amortized over a period of time and from the actuaries comes the table of depreciation. The minute a sponsor lets talent decide all problems in the area of good taste and ethical behavior, he may be letting himself in for trouble.[59]

The discourses surrounding the high price of and increasing creative control by network broadcasting talent were often coupled with discourses on the availability of new television talent. The variety show in particular was often blamed for reifying the vaudevillian headliner and locking out untried comedians. In an article on veteran video comics, George Rosen asked, "What chance has the newcomer, either to survive in the competitive sweepstakes, or to get the opportunity to plant his roots in TV?"[60] A subsequent article in *Variety* claimed, "The major problem in variety talent-buying is the difficulty in getting fresh acts. Once an act has been exhibited at a set figure, price becomes established and it depends on subsequent impact whether it can rise into the upper brackets."[61] Competition between variety programs was apparently a game of topping each other with the biggest names—both in terms of the hosts, as well as guest stars. An unknown act was of little value in this environment. However, some people in the industry argued that with time, television would foster a new generation of comic headliners as "the big black tube [would] become the little red schoolhouse of show business."[62] Although the next generation of talent may have learned from the vaudeo veterans, they had little chance of appearing on their shows. Therefore, if new talent was to develop, it had to find a home in a format that wasn't so reliant on top names to squelch competition.

The increasing cost of top-name talent forced the industry to reconsider its economic and programming structures. If sponsors were to survive financially, they had to produce cheaper programs and leave high-ranked prime-time variety production to the networks or to engage in participation sponsorship. In this way, Weaver was ahead of his time as he first began to give production control to the network and diffuse sponsor investment in comedy programs by rotating sponsors in 1950 with *Colgate Comedy Hour*. By the fall of 1951, all the programs that Weaver considered essential to NBC's position in the ratings were produced by NBC and had multiple sponsorship.[63] As for the other networks, by March 1953, Rosen reported that a revolt was "fomenting among sponsors and agencies over the high cost of television" and many in the industry were seriously considering alternate week sponsorship.[64] Another answer to the rising costs of production as well as the power of vaudeo stars and the dearth of new talent appeared to lie in the production of amateur programs, quiz shows, and sitcoms.

Repackaging the Variety Format

The troubles that surrounded vaudeo stars contributed to a revamping of the variety format. The format was either altered to suit the needs of an amateur program or to include longer sitcom style sketches, thereby fostering little-known performers and lessening the focus on any single vaudeo personality. In addition, presenting genres that differ (however slightly) from the top-name vaudeo programs would appeal to audiences and critics who longed for heterogeneity in programming. Many in the trade press feared that television would have nothing else to offer besides variety shows, anthology dramas, quiz shows, and soap operas:

> Television in the short span of 18 months has practically gone through what it took radio 25 years to exhaust. . . All the tried-and-true formats that have been radio staples for years have been converted into TV—drama, variety, audience participation, etc. Material that had a life-long guarantee in vaudeo and other show biz media has been drained, leaving the TV cupboard threadbare. Result is, everybody is asking, "Where do we go from here?"[65]

The amateur format was certainly not new to broadcasting, as it had been utilized in radio decades earlier. However, it was an alternative to expensive productions such as *Texaco Star Theatre*. "Public desire for variety programs over radio and television has brought about a revival of amateur talent programs unrivaled since the depression days of the early thirties," reported *Variety* in 1950. "[T]he networks have gone out on an amateur talent binge."[66] Citing financial concerns as well as pressures to develop new talent, CBS was especially intent on reviving the amateur format. Most of the network's programs, such as *By Popular Demand* (1950), *The Show Goes On* (1950–1952), and *Prize Performance* (1950) were headed by regular hosts but contained no real stars. The one amateur program that did depend on a big name was *Talent Scouts,* which centered on the personality of Godfrey. NBC also had an amateur lineup with *Ted Mack's Original Amateur Hour* and the short-lived *Lights, Camera, and Action* (1950). These shows certainly brought new talent to the air, but the mark these performers made on the television industry was usually insignificant and short-lived. The most successful amateur hour in terms of presenting talent with any lasting impact was *Talent Scouts,* which introduced performers such as Julius La Rosa, Shari Lewis, June Valli, and the McGuire Sisters.

The variety format was also used in a series of development programs at NBC from 1951 to 1955. These programs utilized talent familiar to only local affiliates as well as formerly unknown performers in various test settings. Soon after CBS's talent raid, NBC vice president John Royal challenged network affiliates to ferret out local talent for the national network in meetings held in 1948:

> There are 178 NBC stations throughout the country and they're all going to be NBC talent scouts, for among the 148,000,000 people in the U.S. that are potential stars it's up to NBC to find them. . . . They can be anywhere and everywhere. Just as baseball has its planned scouting and minor league training . . . just as the film companies are now making a concerted search for talent—it's equally important for NBC.[67]

As discussed in Chapter 2, Royal, along with Warren Wade, did groom local talent through experimenting with an affiliate performance circuit, but it wasn't until a few years later that the finding and training of new talent became a more serious priority at NBC. During fall of 1951, NBC rented out a Broadway theater for a vaudeville-style amateur revue that would serve as a training ground for new talent and material.[68] Designed by NBC president Weaver as an opportunity to test performers' appeal to a live audience (who paid a small admission price to attend performances), "NBC Theatre" was also used occasionally by such established stars as Bob Hope and Jimmy Durante, who wanted to see how their material would play with an audience prior to their on-air appearances.[69] The production was a part of a larger plan outlined by Weaver in a November 1951 memo that began with a description of Weaver's goal: "The broad concept of the comedy development plan is based upon the fact that: (a) comedy is not only desirable, but essential to television programming and that (b) there is virtually no place today where young comedians, comedy writers and directors have an opportunity to develop."[70] The memo announced the hiring of writer/producer Joe Bigelow who would work as a talent scout for the network, looking for unknowns as well as convincing successful comedians—such as Jack Paar—to enter into television. He also was to work with comedians who proved to be only middling on television, but still seemed to have potential (the memo points to Henny Youngman, Bert Lahr, and Red Buttons among others). The network would provide other options for new and developing talent including: performing as a warmup act for a studio audience before a television show; performing on *NBC-Tryouts,* a two-hour Saturday afternoon amateur program that would be broadcast from three different cites—New York, Chicago, and Los Angeles; or an opportunity to perform at NBC Theatre.

The following winter, NBC initiated another phase of its comedy development plan with a series of forty-five-minute audition "showcases" held on alternate Thursdays in a large studio.[71] Most of the individuals who performed had never worked as a headlining act but had perhaps appeared in small clubs or vaudeville houses. Talent was allowed rehearsal time in the studio and was sometimes helped out with their material by network staff writers. By the following year, the director of the program, Bill Gargan, had found at least fifteen comics to appear on a broadcast version of the showcases that would air Sunday nights on WNBT. Because, according to Gargan, "the important thing about developing new comics is the need for exhibition," only one comic and a singer were spotlighted each week.[72]

In 1955 Weaver began the third phase of the comedy plan. After performing in clubs and local theaters, select comedians would be showcased in a local show on KRCA called *Komedy Kapers.*[73] The most prominent names who went on to work for the network after their appearance on *Kapers* were Jonathan Winters, George Gobel, Kaye Ballard, Pat Carroll, Mort Saul, and Shecky Greene.[74] One November 9, 1955, memo from Fred Wile, director of the development program, to NBC's general program executive for the Pacific division, reiterated the goals of the program: "Our strategy is to achieve this development [of comedy stars and writers] by means of showcasing and providing exposure under battle conditions to comics, already established but not yet all the way "there"—to young comics, to test ideas for spots and vehicles in vignette form. . . . IF we develop one Gobel or one future Oppenheimer or Josefsberg or Mankiewitz, there will be enough credit for everyone—to say nothing of satisfaction."[75] But many advertisers would have disagreed that one new face was enough, particularly because television seemed to be able to go through them so quickly. A 1955 issue of *Tide*, a magazine for the advertising and sales industry, polled five thousand "leaders in advertising, public relations, sales and related fields" on the subject of the shelf-life of television stardom.[76] Most of them responded that an increase in "viewer fatigue" and "fickleness" when it came to television performers was becoming a real problem for television—one that was unlikely to go away. Whereas 80 percent of the panel agreed that George Gobel was television's "hottest comedy find," they also predicted that the star would not survive more than two years as a viewer attraction. Interestingly, 63 percent of them thought that the variety format "most severely limits a stars life," believing the sitcom far more conducive to helping performers retain their popularity.[77]

Although amateur and other alternatives contributed an element of difference to the variety format and enabled sponsors and networks to exploit the format's popularity without the expense of stars, producers of traditional vaudeo programs such as *Your Show of Shows* and *Texaco Star Theatre* were adding characters and story lines in order to deemphasize vaudeo's urban sensibility and reliance on guest stars. Additionally, especially pertinent after the establishment of the 1952 Television Code, such preplanned storylines promised producers more control over vaudeo's questionable content.

Texaco was the first to undergo such a transformation, incorporating regular storylines and eliminating much of Berle's direct address to the audience. The veteran radio writer Goodman Ace was hired to help reconstruct the program to include more plot. New cast members were brought in to play Berle's agent, secretary, and stagehand. In one memorable episode guest starring Gertrude Berg, Berle's act was spliced together with the narrative comedy of *The Goldbergs* as Molly Goldberg was scripted to be a matchmaker for Berle and his secretary. Berle was able to utilize his on-stage persona as the lengthy sitcom sketch was written as to take place in Berle's off-stage life. In enabling Berle to retain his persona, the new "backstage" narrative of the program retained characteristics of *Texaco*'s original format while simultaneously trying to reap the benefits of the episodic narrative. Despite such alterations, however, Texaco dropped Berle's program at the end of the 1952–1953 season, and the show continued in its revised format through 1954 as the *Buick-Berle Show*. George Burns explained why Berle's show dropped in ratings even after it instituted a new family-friendly format: "After five years, Milton's audience had seen everything in his closet. He tried to change, but then he wasn't Milton. I mean after you've seen a man wearing a dress and a wig, with two blacked-out front teeth, standing in a tank of water, getting hit in the face with a powder puff while a wind machine blows whipped cream all over the stage—four times—everything else is a little anti-climactic."[78] A 1952 *Variety* review of the revamped show contended that Berle's program was simply no longer funny:

> Somewhere along about the middle of the "Texaco Star Theatre" pre-miere last Tuesday night (September 16), a befuddled Milton Berle commented, "Whatever became of the Berle show?" It was one of the most apt cracks on the program. For somewhere along the line in the widely-touted attempt to inject some adrenalin into the slipping Texaco stanza, they forgot to make Berle funny. . . . The result was an entirely different but somewhat incongruous Berle—a Berle who for practically a full hour found himself playing straight man. . . . If the intent was to throw the old Berle out the window and substitute a restrained, sympa-thetic fall guy for the others, it succeeded.[79]

The sitcom-within-a-variety-show strategy had been used successfully in Benny's radio and television programs as well as CBS's popular *Burns and Allen Show*, which had begun its run in 1950. Both Burns and Allen played themselves on the show, yet it was Burns who played omniscient narrator and host, able to cross the diegetic "fourth wall" from vaudeville stage to sitcom set. Although the characters that Burns and Allen played were obviously an act rather than a direct representation of their "authen-tic" personalities, their program was constructed in such a manner as to give the audience the impression that it was getting a glimpse of the couple's domestic life. *Burns and Allen's* self-referentiality, blend of reality and fiction, and complicated wordplay made it the most well-crafted model of television's blend of sitcom and variety formats.

Your Show of Shows was initially conceived as a series of sketches high-lighting Caesar and Coca's comic skills, which was broken up by mono-logues and production numbers. But in 1953, the show jumped on the sitcom bandwagon and incorporated fewer, but longer, sketches and made sure to place at least one of these sketches in a domestic setting. After its first successful year on CBS, *The Jackie Gleason Show* (1952–1959) found itself straddling formats as well. The first half-hour of the program was filled with production numbers by the June Taylor dancers, Gleason's monologue, and sketches based on characters that Gleason had made famous on *Cavalcade of Stars* (1949–1952, Dumont) a few years earlier.

Fig. 4.3 Gleason and his *Honeymooners* co-star Art Carney.

One sketch, "The Honeymooners," which appeared to be the favorite of Gleason's audience, eventually took over the entire second half-hour of the program. By 1955, the show would temporarily dispense with its variety format, transforming itself into a half-hour telefilmed sitcom now titled *The Honeymooners*. During the program's final season, *The Honeymooners* returned to its former status as a sketch within the show's hour variety format.

The expansion of sketches and the hiring of additional cast members and writers dispersed creative input and audience attention across a number of individuals. The vaudeo star was no longer the sole point of audience identification and producers sustained greater authority over program content and the scripting of commercial messages. Recurring characters and settings helped standardize production practices and enable the program to rely on plot and character to attract an audience, rather than being completely dependent on the personality of a sole vaudeo comic. In addition, expensive Hollywood guest stars were no longer a key component of a program's success. The introduction of telefilm would further standardize these programs and take much of the strain off of its talent.

The variety-sitcom blend was a precursor to the domestic comedy that came to dominate network schedules by the mid- to late 1950s. CBS, always looking for a way to topple NBC from its number one ranking, was the first to invest in this type of format. "With NBC having cornered most of the top-name comedians for its Wednesday and Sunday night television shows, CBS is turning to situation comedy as its answer to the NBC talent line-up," began a front-page article in a February 1951 edition of *Variety*.[80] As a result seven new half-hour domestic comedies appeared on the network's schedule in the fall of 1952 including *Our Miss Brooks*, *Life with Luigi*, and *My Friend Irma*.[81] The proliferation of comedies of this ilk led *Variety* to proclaim, "That television is on an 'Everybody Loves Lucy' kick is evidenced from the 1952–1953 programming rosters. In contrast to the handful of situation comedies circuiting the video channels last season, they're now all over the spectrum."[82]

Spigel argues that situation comedies such as *I Love Lucy* (1951–1957, CBS) and *I Married Joan* (1952–1955, NBC) were a result of a cultural negotiation of past and present values. She writes, "These programs allowed people to enjoy the rowdy, ethnic, and often sexually suggestive antics of variety show clowns by packaging their outlandishness in middle-class codes of respectability."[83] Ironically, many of the "new" stars of these programs were not, as the industry had promised, unique to television. Lucille Ball was a former film actress and radio comedian. Her talents were not honed on the vaudeville circuit like many of television's earliest stars, but her performance style did incorporate the physical humor and sight-gags of the vaudeville aesthetic. Television would develop a number of stars of its own, but would still remain dependent on other entertainment industries for the majority of its talent.

The 1951–1952 season proved to be a major turning point for the television industry and had significant implications for the form and functions of television stardom. In his analysis of the combined Trendex-American Research Bureau's ratings at the end of the 1951–1952 season, Rosen drew five major conclusions: (1) the "Milton Berle Era of TV leadership" had come to an end; (2) the sitcom formula as expressed in CBS shows such as *I Love Lucy* was the next big trend in programming; (3) NBC's multimillion dollar investment in "top comedy personalities and rotating comedy programming" appeared to be misguided; (4) Godfrey was still a valuable television commodity; and (5) the decline of the *Red Skelton Show* "invites speculation as to the wisdom of negotiating [talent] pacts extending of many years."[84] In retrospect, Rosen's analysis was spot on. The industry was changing and the norms of comedy performance and genre were about to go through a significant alteration. However, there were consistencies that would carry through, at the very least, the next few seasons, including the role of the television comedian as product salesperson and, often, the embodiment of a network's commercial imperatives. Godfrey, who outlasted the trials of the 1951–1952 season, was the ultimate television performer in both respects. In the following chapter, his career and its relationship to the larger discourses of commercialism and authenticity that surrounded broadcast stars will be discussed in greater detail.

CHAPTER 5
OUR MAN GODFREY
Product Pitching and the Meaning of Authenticity

Obviously, a television performer had to have considerable talent and a great rapport with his audience in order to survive on television. But he also had to have the ability, and motivation, to sell. A performer's career could end rather quickly if he or she did not move enough product. (Sid Caesar was a notable exception to this rule, as he refused to present products or ads in *Your Show of Shows*.) As Hal Davis, the promotion vice president of Kenyon and Eckhardt advertising, warned in 1954, "People will buy products pushed by personalities they like," and stars who "refused to deliver commercials won't be around long."[1] It was generally understood that a television star's image had to be consistent with the image of the sponsor's product in order to successfully initiate viewer identification with and desire to purchase a specific brand. In addition, a star had to exude an honesty or "naturalness" that would engender trust in the audience. George Burns acknowledged the dual responsibility a star had when acting as a product spokesperson: "[Gracie and I] don't try to kid people, but we never forget we're supposed to sell Carnation milk. We make every effort to do it as honestly as possible. If we don't sell the product, we don't have a show."[2]

Career longevity was not a concern during the immediate postwar era for Arthur Godfrey, who was consistently cited by radio and television advertisers as one of their favorite "pitchmen."[3] This was because, in his role as a talk and variety show host, Godfrey had an unprecedented hold

Fig. 5.1 Arthur Godfrey pushing Pillsbury flour and Chesterfield cigarettes. Library of American Broadcasting.

over postwar consumers. In a 1956 letter to *The Saturday Evening Post*, Mrs. Anthony Firley voiced a typical sentiment of Godfrey's fans when she wrote that she and her husband maintained "the deepest respect and admiration for Mr. Godfrey and bought $1,195.75 worth of Frigidaire appliances last year to prove that we believe in Arthur's sincerity where 'his' products are concerned; to us when Arthur says 'I know it's good' it's the seal of approval and if we need it we'll give it preference over ANY other brand."[4] A 1950 article in *Sponsor* noted, "Godfrey's commercials are generally conceded by the public and the trade to be both entertaining and credible. This latter quality, of course, springs largely from the effect of sincerity possessed by Godfrey in so remarkable a degree."[5] Dozens of popular and trade press articles during the late 1940s and early 1950s acknowledged Godfrey's unique ability to generate unprecedented revenue for CBS and sponsors. Much fuss was made over Godfrey's $440,500 salary, which made him the highest-paid performer at CBS in 1948,[6] as well as the fact that, during his years on television, Godfrey brought somewhere between $11 million and $20 million to the network (approximately 12 percent of its overall revenues).[7] Consequently, the popular press consistently and explicitly connected Godfrey's public image with financial success. For example, the February 27, 1950, cover of *Time* pictured Godfrey dressed in a Hawaiian shirt and straw hat speaking into a microphone shaped like a dollar bill accompanied by the seemingly incongruous caption, "Arthur Godfrey: He has empathy." The connection between empathy and a dollar sign made in this cover was an acknowledgment of how Godfrey's success with sponsors and audiences was due largely to his ability to project sincerity and genuineness in his role as host and sales pitchman in his television and radio programs.

Although this combination of characteristics worked well for Godfrey for many years, his career suffered a major blow in the fall of 1953. On the morning of October 19, Godfrey fired Julius LaRosa (arguably one of the most popular singers ever to arise out of Godfrey's cadre of infamous "friends") on the air. Before LaRosa's spot on that morning's *Arthur Godfrey Time,* Godfrey surprised the singer by asking him if he thought doing the show "was a pain in the neck." Although reportedly confused by the comment, LaRosa went on to perform his song as planned. Yet once LaRosa concluded his number, Godfrey unexpectedly told the audience, "That, folks, was Julie's swan song." This resulted in a public relations nightmare for Godfrey. According to press reports that followed, Godfrey was angered by what he perceived as LaRosa's "cockiness," which had arisen because of the singer's awareness of his rising popularity with the show's audience. (It was said that LaRosa's fan mail had begun to outnumber Godfrey's reported sixty thousand pieces of mail per month.) Godfrey also explained that he fired the singer because he had been late to dance rehearsal in the weeks leading up to the October 19 broadcast and that he had hired an agent—something Godfrey had forbidden his "friends" to do. But it was the reason Godfrey gave in a press conference that "Julius had lost his humility" that startled reporters and fans, setting

Fig. 5.2   Arthur Godfrey with Julius LaRosa. Library of American Broadcasting.

off a series of critical magazine articles and angry fan letters. Although Godfrey retained many die-hard supporters, many other people were disturbed by what they perceived as cold-heartedness. It was one thing to let a cast member go, but to fire a beloved and unsuspecting "friend" in such a public and humiliating manner seemed unnecessarily cruel. The resulting uproar led the *LA Examiner* to muse, "Is the 'lovable redhead' only a myth now coming apart at the seams?"[8]

In this chapter, I will use the construction of Godfrey's persona and its virtual dissolution in 1953 as a way to talk about the role that the notion of authenticity played in postwar television's complicated and decentralized negotiation of stardom. Godfrey's case is fascinating for a number of reasons, but the issue of the "authenticity" of his persona, and its relationship to his ability to pitch product, is one that is central to the larger delineation of postwar television stardom. Throughout his broadcasting career, Godfrey and his press agents carefully exploited his image as a genuine and common man. Yet this persona had to be weaved through the numerous subjectivities held by Godfrey: that of a salesman, a talent scout, a variety performer, a family man, a military and aviation expert, and a man who *appeared* untainted by the machinations of celebrity, wealth, and (ironically) the commercial trappings of the broadcast industry. All of this came apart when the framing of LaRosa's firing showed the public that he also could be a ruthless, ego-centered businessman. The press and certain factions of the audience began to attack his original persona as just an act.

The Man with the Barefoot Voice

As a radio announcer and television talk and variety show host, Godfrey fit a fairly typical profile of a television performer of the period except that, as he explicitly (and somewhat jokingly) acknowledged, he had no "real" talent. After working as a radio technician and announcer in the Navy, Godfrey began his broadcast career as announcer for NBC's Washington, D.C., affiliate in 1930. Initially, he used a fake British accent on the air, which he dropped when he realized that the formality of his voice was coming between him and his audience. He also claimed to come into his signature pitch style in a 1934 broadcast when he was asked to deliver a commercial for a Washington department store that was trying to sell women's lingerie. Godfrey's voice rose in pitch as he read the copy, which described the product as "filmy, clingy, alluring silk underwear in devastating pink and black." He then whistled and whispered to the audience, "Is MY face red!"[9] Godfrey claimed that his bosses threatened to fire him for this ad-lib until they learned the department store had sold out their stock of lingerie within hours of the commercial's airing.

By 1934, Godfrey had moved over to CBS, acquiring a large following not only in Washington but also New York. His crossover into television came in 1948 when he accepted the role of host of CBS's talent showcase *Talent Scouts* (1948–1958). Unlike comedians such as Jack Benny and Burns and Allen, who had to alter their radio programs and performance styles considerably for television, Godfrey made few changes to his format or style on entering the new medium. Some of his programs were simulcast and he spoke in a relaxed, slow manner to his studio and home audiences. Believing that the key to radio was to perform as if speaking to a single listener, Godfrey became known for what Fred Allen called his "barefoot voice," which was predicated on a "naturalness" of delivery and use of common lingo such as "ain't" and "doggone," which stood in direct opposition to most announcers who spoke formally and seriously.[10]

By 1952, Godfrey was producing ten hours of live daytime and prime-time programming a week. In addition to *Talent Scouts*, he starred in *Arthur Godfrey and His Friends* (1949–1959), his Wednesday night variety show, *Arthur Godfrey and His Ukulele* (1950),[11] and *Arthur Godfrey Time* (1952–1959), his daily radio morning program, all of which ran simultaneously on CBS and were said to attract a weekly audience of over forty million.[12] His approach in his programs was to enmesh the project of selling into his persona—a move that many top performers were reticent to make. Yet Godfrey was not the type of overenthusiastic spokesperson who one might find in certain postwar television commercials, as part of Godfrey's "shtick" was to ad-lib during these spots, often openly criticizing and joking about many of the claims that advertising agencies made for their clients' products. This strategy, although occasionally off-putting to agencies and sponsors, enabled Godfrey to appear more believable to consumers than someone willing to just mouth scripted endorsements.

Writers in the popular and trade presses endlessly ruminated over the traits that led Godfrey to such prosperity.[13] In the *McCall's* article "Why Women Love Arthur Godfrey," Isabella Taves explained that Godfrey "has a combination of small-boy naughtiness and genuine sweetness which adds up to sex appeal. He teases 'the goils' in his audience. But when he decides it is time to get in a good hard sell for one of his sponsored products he drops the foolishness and says winningly: 'My dear, why don't you go out and buy a box of Rinso today? You won't be sorry.'"[14] It was also reported that one woman wrote to Godfrey thanking him for giving her what her own family could not, "My husband and my children ignore me. You are the only person in the world who really talks to me."[15] Writing in *Cosmopolitan*, Joe McCarthy argued that it was not only Godfrey's sex appeal, naturalness, and "solid sense of values" that made him so popular with audiences and sponsors but also the intimacy he shared with his audience, his willingness to do anything in front of a television camera, and the unpredictable nature of his advertising plugs.[16] McCarthy recalled a particular installment of *Arthur Godfrey's Talent Scouts*, wherein Godfrey was honest about how he felt at that moment about his sponsor's product:

Godfrey plays his Chesterfield-cigarette commercials fairly straight. But on one Wednesday-night show, the commercial came after Arthur had done a strenuous ice-skating number. Still breathing heavily, he attempted the Chesterfield commercial. He looked at a cigarette, grimaced, and threw it away. "There are times, " he told his audience, "when you just don't feel like smoking."[17]

Godfrey's sponsors, which included Chesterfield, Lipton, Pillsbury, Nabisco, and Lever Brothers, were tolerant of such moments as they seemed to give credibility to Godfrey's eventual endorsement of their products. Indeed, they were working under the belief that "[T]he best commercial you can give Arthur is to put the product in his hands."[18]

Ernest Dichter believed that Godfrey's joking about his sponsor's products was the central factor in his believability as a product spokesperson: "Godfrey behaves the way you think you would have to if you were in his shoes. The normal person feels that if he were working in the advertising field, he would have to make fun of it once in a while. That he would not possibly be able to take it with a straight face, day in and day out."[19] A 1953 *Collier's* article also stated that the audience's trust in Godfrey stemmed from his ability to "kid the commercials," arguing that "people feel that since he doesn't try to oversell his sponsor's products, he must be honest about the products and about everything else."[20] Godfrey himself reiterated this belief and felt that it transferred to his overall reception as an authentic individual: "[The audience knows] that for twenty-three years I've never told them anything but the truth about the product, so they feel I'm on the level with them when I talk about other things."[21] Fans were so willing to buy whatever Godfrey endorsed that his references to and personal use of ukuleles, Hawaiian shirts, aviation, and travel to beaches in Miami Beach and Hawaii led to marked sales increases for these industries and locales. Godfrey was even credited with being a major factor in the rise of airline travel during the 1950s, because if he felt that flying was safe, so did significant numbers of consumers. Godfrey was quoted as saying, "I keep selling my fans on the fact that no airplane ever had an accident that couldn't be traced to human error. Airplanes don't kill anyone. Just man kills man."[22] (The aviation industry actually rewarded Godfrey such unpaid plugs by presenting him with his own customized DC-3.)

Nonetheless, the sarcasm Godfrey deployed in his commercials was clearly an important point of identification for his viewers. Complicating perceptions of the postwar consumer as one that wholly embraced the trappings of commercial culture, Godfrey's viewers responded empathetically to

Fig. 5.3   Godfrey demonstrates the cleaning power of Rinso. Library of American Broadcasting.

his disavowal of blind participation in commercial process (even as they were purchasing many of the products he recommended.) Godfrey's play and critical self-awareness also helped negotiate some of the discursive tension revolving around broadcast commercialism. It is important to note that the decision to make broadcasting a commercial rather than a public entity at its inception had been reasserted and validated continually through ongoing negotiations between industry agents and governmental institutions. As Thomas Streeter points out, commercial broadcasting was (and still is) a "product of deliberate political activity" that resulted in "favoring some people and values at the expense of others."[23] This validation process also worked to create audiences who generally accepted broadcasting's commercialization (although the extent and manner of its implementation continued to be at issue in public discourse) and who understood that programming was being directed to them as consumers. Television stars in general, in their roles as product pitchpersons and embodiments of

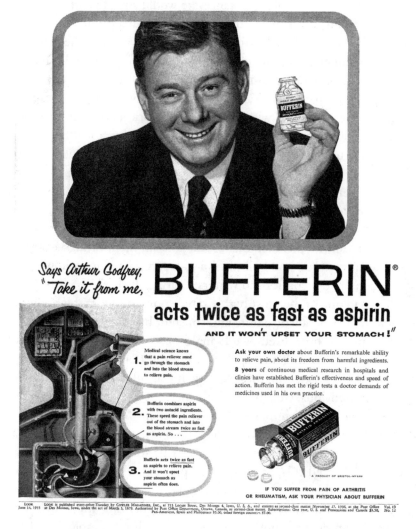

Fig. 5.4 Godfrey for Bufferin in 1955.

television's economic and narrative imperatives, helped negotiate and give voice to the medium's private interests providing the audience with human and, perhaps, "sincere" representatives of broadcast commercialism.

Although educators, politicians, and other arbiters of culture concerned themselves with the effects of television sponsorship, and advertising agencies became increasingly powerful producers of ideology, Godfrey offered a unique middle ground: a discursive space in which advertisers and their claims could be processed by a stand-in for the American postwar consumer. At times, Godfrey would deflate advertising's puffery or disparage specific claims, whereas in other instances he would stray from the copy he was given to recount his own positive experience with a product or service. By inserting his opinion in advertising copy he intervened in the sponsor's address to consumers, thereby temporarily disrupting and mediating the commercial aims of his program, and ultimately succeeding in getting his viewers to purchase particular wares. Furthermore, when coupled with his performance style and bound within the larger rhetorical positioning of television realism, his pitch style also helped further his claims to authenticity.

Godfrey was not the first broadcast star to employ spontaneous cynicism in his plugs. Fred Allen prefigured Godfrey's negotiation of broadcasting's commercialism and was even more voracious in his criticism of American commercial culture. As discussed in Chapter 1, Allen acted as perpetual outsider, attacking almost every aspect of the industry in which he worked. His shows included bits on vacuous ad men, ratings-obsessed network executives, vain and hollow film stars, annoying quiz show hosts, and beleaguered radio comics and writers. During the early to mid-1940s, when Allen had one of the most highly rated shows on radio, his audience found pleasure in his caustic nature and constant barbs at the network system that employed him. Fred MacDonald observes, "To an audience of 'average' listeners earning 'average' salaries, and possessing an 'average' share of powerlessness within mass society, it was often refreshing to assail in jokes the pretensions, wealth, and influence of social leaders. No comedian better exploited this impulse than Fred Allen."[24]

Allen embodied the same industrial paradox that Godfrey did: he was an avid critic of commercialism, yet he was still able to move a sponsor's product. He achieved this, in part, by employing self-referential humor that poked fun at the sponsor's product, without denigrating it. Allen also satirized radio commercials, mimicking and attacking their form and content while not ever naming any real sponsors. For example, in a 1938 program Allen playfully chided a made-up meatball product, the primary selling point of which was its bounciness:

> Mother Murphy's Meatball: If you drop a Mother Murphy's Meatball on the floor, don't be embarrassed. It will bounce back on your plate again. Have fun with your meatballs, folks. . .Dribble them up and down the table. Tie a meatball on your spaghetti and make a yo-yo. But buy them. Buy—spelled B-U-Y.[25]

Yet NBC was much more upset with Allen over his barbs than CBS was with Godfrey and his digressions. One reason for this was that Allen was relentless in his attacks on networks and advertising agencies as inept and hypocritical cultural gatekeepers. He used the industry as a comedic scapegoat for the stifling of creativity and the commercial polluting of the airwaves, thereby providing an outlet for criticism and resistance by the populous. Still, he managed to increase not only his ratings but also the sales of his sponsors' wares. This reflects Michele Hilmes's assertion that Allen's show "resulted in a carefully negotiated balancing act that directly reflected and commented on the position of radio itself."[26]

Alan Havig argues that Allen's cynicism dovetailed with the 1930s mass audience's suspicion of big business and commercial culture.[27] Whereas Godfrey continued Allen's tradition of ad-libbing and skepticism in his advertising plugs, his approach was not as strident or as biting as Allen's. Furthermore, whereas Allen presented himself simultaneously as a populist and an intellectual, Godfrey aligned himself thoroughly with the average American. Allen would probably have bristled at being compared to Godfrey since, in a 1950 interview, Allen railed against him as representative of a downward trend in broadcasting, saying:

> [Godfrey's] sweeping the country, and, Lord knows, it needs to be swept. But I think Arthur must be doing it with a short-handled broom—he's nearer to the dirt than most people. Millions of people think he's the funniest guy alive, but their standards are open to question. This is an age of mediocrity. Anything mediocre is bound to be a success. As we get more regimented, there are few Tiffany's and more Woolworths.[28]

Yet most commentators and industry insiders interpreted Godfrey's ability to project ordinariness not as mediocrity but as a strength. According to them, this feature facilitated audience identification and pleasure. William Paley, who was CBS board chairman at the time and who based his programming philosophy on the creation of a network star system, argued that Godfrey "is the type of guy the average man would like to be. He's a wistful projection of the average guy."[29] Joe McCarthy quoted an unnamed industry observer as saying, "The real secret of Godfrey's success is that while he talks about so many things and does so many things he still remains like a guy who lives across the street. There is nothing theatrical about him. Compared to him even Bing Crosby is a little stagy. Godfrey is the first entertainer since Will Rogers who seems free of make-believe."[30] And an eight-page article in *TV-Mirror* describes Godfrey's everyman status as an achievement of shape-shifting of sorts, as the author wrote that the star "might be the neighbor you'd least mind lending your lawnmower to. He might be anyone in your acquaintance: the friendly insurance man, that genial short order cook down at the corner beanery, that nice chap you worked with on the assembly line, that truck driver or door-to-door canvasser or farmer—guys you've known."[31] Even though Godfrey would often publicly discuss his lavish lifestyle (travel to foreign locales, his Virginia horse farm, his stable of airplanes) and exorbitant salary, he still managed to portray himself as an "everyday Joe." Specifically, Godfrey was said to eschew the "glamour boy" trappings of celebrity life. [32] He lived in a "unpretentious two-room suite in a medium-priced hotel" when working in New York, avoided showbusiness haunts, went to local barbers, and wore some suits that were "five years old."[33] Godfrey's extratextual material encouraged audience identification with him as "a man next door, a neighbor and a friend, the average man made good."[34] Godfrey was raised in a struggling New Jersey family, and it was widely reported that he had to quit school at a young age to help support his family. Descriptions of these meager beginnings along with his well-publicized service in the military certainly helped create the image of Godfrey as an ideal representative of the American dream, and his off-screen adventures helped to further the popular belief in the good life of postwar prosperity. Surprisingly, Godfrey's wife, Mary, was rarely mentioned by name in such coverage, but his three children were often discussed in relation to Godfrey's desire for normality. Supposedly, the children, who lived on the farm in Virginia, had to work hard for their "modest" allowances, ride the bus to public school, and were shielded from publicity. "They're nice normal American kids," said Godfrey. "They aren't interested in show business, and they only watch television to see what their old man is going to do next. I don't care what they do when they grow up, only I hope it isn't show business. I'm too old to quit now. But I don't want them in it."[35] This comment, a strategic public relations move, spoke to contemporary discourses on the negative effects of television on children while making the highly dubious assertion that Godfrey really didn't care for his own career choice. The idea that Godfrey was "stuck" in an industry he felt he needed to protect his family from (even as it provided them with a more than comfortable lifestyle) resonates with, and ultimately underscores, his on-air role as a mitigator of the commercial broadcast industry's attempts to reach or perhaps exploit the American consumer.

Intimacy and Authenticity

Beyond the extratextual construction of him as an ordinary American, Godfrey's on-air spontaneity and speaking style further positioned him as genuine and sincere. In 1947, Paul Whiteman of *Variety* argued that one of the major attractions of television would be its "fresh unrehearsed naturalness" and that "it will be the ideal field for boys and girls who are talented ad libbers."[36] Although Godfrey could not sing, dance, act, or play the ukulele very well, he was an expert at ad-libbing and rarely used scripted material. He also smoothly weaved his audience into his daily morning program by directly addressing the studio and home viewer and by making the reading of his fan mail a regular routine. Yet, the authenticity of his persona was also a product of larger discourses on the inherent aesthetic qualities of and social role played by television during these years.

Current star theory posits a spectator's search for an authentic core to the star image/persona as a central tension of star spectatorship. A star is understood by a spectator to exist outside the media text—as a living, private, human individual. In recognizing the constructed nature of the media image, a spectator scours textual and extratextual information for clues to the "real" identity of a star. This theory is discussed by Richard Dyer, and others, exclusively with regard to film stars. Although it seems a similar play surrounding the authentic existed (and exists) between early broadcast stars and audience members, the boundaries between what is perceived to be authentic or constructed were quite different.[37]

Television's claim to intimacy and immediacy both with regard to its overall presentation and the personality of its performers leads into the way in which television stars were constructed and received. Specifically, broadcasting economic goals and its presumed aesthetics encouraged intimacy in television performance style along with an enhanced conflation of a star's "real life" with that of his or her character's textual history and personality in order to promote viewer identification with both a program's star and product(s). As Dyer and Richard deCordova have discussed, the search for authenticity in film star texts required the authentic to be held back from the spectator by the ultimate inaccessibility of the star's "true" identity.[38] However, within the context of the discourses that constructed television's aesthetic, television viewers were encouraged to believe that they could actually locate the true personality of a television star somewhere within his or her performance. In fact, this belief might have been stimulated by the popular rhetoric on television's intrinsic aesthetics and the "naturalness" from its spontaneity that television performers displayed.

In other words, because television was said to produce intimacy, immediacy, and spontaneity, it also generated authentic identities. Whereas film encourages spectators to pursue the "real" in the movie star, the discourse of television aesthetics appears to thrust the authentic on its audience.

This notion of television's unique capacity for verisimilitude also informed the way in which performance on television was constructed and received. As William Eddy asserted in 1945, "The actor must always remember that the main factor of all successful television productions is the element of naturalness and he must work toward that end. The closer that the actor, by reason of his experience and knowledge of the system, can approach this goal, the higher will rise his star in the new firmament of the video arts."[39] Thomas Hutchinson, one of the first people to teach television production at New York University, reiterated this sentiment, suggesting in 1946 that "[t]elevision brings the man himself, his personality, and his charm. . . . The right performer with the right personality and the right material is a television natural."[40] In the rhetoric of the immediate postwar era, television also was thought to require a natural, genuine, or *authentic* performance style. This rhetoric often implied that television would help to reveal the true personality of a person who performed on its screen. In a 1951 *Variety* editorial, Eddie Cantor explained that "television is murder on the phony. Those brutal cameras, those revealing close-ups, are tougher than the Kefauver committee. TV exposes hypocrisy, insincerity, anything that's fake and dishonest. That television screen in the living room tells you more about a man's insides than the x-ray machine in a doctor's office."[41] Donald Curtis used similar imagery when he wrote, "The television camera goes inside of an actor's mind and soul, and sends the receiving set exactly what it sees there."[42]

Nevertheless, the authentic was still obfuscated or dislocated by a layering of star subjectivities. As deCordova's work has shown, the concealment/revealment of cinematic star discourse was in part an industrial marketing strategy aimed at getting the moviegoer back into the theater to see or discover more about a particular star or to "extend the boundaries of the cinema as institution so that it could more fully occupy people's lives."[43] This being the case, the drive to keep the television viewer would be conceived somewhat differently than the cinematic process deCordova outlines. Perhaps the goal was to naturalize the performers in the domestic setting so as to make them appear less aberrant in the context of the everyday, yet to simultaneously make them engaging enough to capture the audience within the continuing story and to draw positive attention to

Fig. 5.5 Godfrey for Pepsodent.

the sponsor's product. Yet, less of an industrial drive existed to get the television viewer to return to his or her set as exists a need to have him or her return to a particular program and its surrounding product(s).

Television stars of this period maintained multifarious identities. They often were required to act simultaneously as a program's host, a spokesperson for the sponsor's product, a representative of a network, one or more characters, a public personality, and a private individual. And all of these roles sustained various interpolating positions along the spectrum of extraordinary and ordinary. As Denise Mann has pointed out, television stars had to be grand, but they also had to naturalize themselves into the domestic domain while mobilizing their personas smoothly and obliquely to move product.[44] These roles required the star to simultaneously remain extraordinary while asserting his or her ordinariness. This not only contributed to the audience's fascination with them as individuals but also furthered the process of identification with regard to the products sold by such stars.

Advertisers tried to achieve a meshing of the "personality" of their product with that of the star/character's image. Motivation research (such as the work done by Ernest Dichter) was an essential tool for advertisers and broadcasters during this period. Most of the results of this psychologically based research emphasized the effectiveness of associating a product with a "realistic" television personality. For example, the Weiss & Geller advertising agency put together a panel of eight social scientists ("two psychoanalysts, a cultural anthropologist, a social psychologist, two sociologists and two professors of social science") to watch prime-time programming in order to analyze the most effective personality type to deliver commercials.[45] They reported their results in a May 1954 article in *Broadcasting, Telecasting,* concluding that television advertising's ideal personalities "are those who present themselves as human and fallible, who by their identification with consumers induce the consumer to identify himself with them. . . . Such personality is reassuringly like you or me, or like our husbands or wives. And the product becomes more believably identified with real human needs."[46] Indeed, Godfrey fit this model with his relaxed delivery and performance style, but the regular cast of performers he surrounded himself with also contributed to the construction of a naturalized on-air domestic setting. The Weiss & Geller study analyzing the way in which the host's personality influenced commercial messages found that Arthur Godfrey's morning show, *Arthur Godfrey Time,* represented family relationships and a home environment without the use of props. The panel concluded:

> Psychologically, his morning program creates the illusion of the family structure. . . . There is no mother in the Godfrey family. This gives the house-wife-viewer the opportunity to fill that role. In her fantasy, Godfrey comes into her home as an extra member of her family, and she fancies herself a specially invited member of his family. She entertains him in her home, but at the same time, he invites her into his studio.[47]

Commentators later theorized that the firing of LaRosa represented a breakup of this on-air family, and upset the program's largely female audience because it appeared that patriarch Godfrey kicked out one of the program's "sons" without consulting the rest of the family or viewers. The press took LaRosa's firing as an opportunity to interview people who had worked with Godfrey in an attempt to expose the "truth" behind Godfrey's construction. An unnamed ad agency executive who represented one of Godfrey's sponsors told the *LA Examiner* that "Godfrey is an utter dictator. Disagreement is verboten . . . Basically he's a very embittered man. He doesn't know the meaning of humility."[48] Another anonymous source claimed, "Until a minute before airtime, Godfrey is a ruthless tyrant. A minute later, on the air, he becomes a charming Huck Finn. A minute after the show ends, he sheds his stage personality like a cloak and rushes off in a rented limousine, without a word."[49] Godfrey responded to such accusations by blaming the industry's morally dubious norms for their contempt for his rebelliousness and for his refusal to conform, "Advertising people may hate me because I've been rough with the jerks who write the commercials. . . . Press agents sneaking around here for some phony deal? Yeah, they hate my guts. Song pluggers? Yes, they'll hate me because I insist on picking songs myself. Maybe someone got lost in the shuffle. . . . Maybe in a busy day you step all over a lot of people without knowing it. But it was never done intentionally."[50] Yet, he contended, at least for a while, that his fans still adored and trusted him.

Although some fans did continue to support and believe in Godfrey's original persona, others were put off by what they perceived as his inherent insincerity. John Crosby wrote a lengthy article for *Collier's* in late October 1955 contending that it was the many years of adulation that Godfrey received that turned the performer into an all-too-powerful narcissist.[51] The letters sent into *Collier's* after the publication of Crosby's article reveal the voracious response that the public had to the scandal. Although some people wrote to *Collier's* agreeing with the author, many were angered by Crosby's claims that Godfrey was not who he appeared to be. In a letter to the editor, Joyce Liberman of Pittsburgh wrote, "John Crosby is a fat, dissipated old critic who probably never accomplished anything truly worthwhile in his life. Whereas Arthur Godfrey is virile, strong, competent, resourceful and is always doing something for the country he loves. It is clear to see that Crosby is burningly jealous of Arthur, who is a REAL man."[52] Arline Koogler, of Hawthorne, California, would have agreed with such a sentiment, as she claimed, "I know thousands of women who would be more than happy to beat John Crosby to death with tea bags. Arthur Godfrey is an ever-lovin' doll, sirs."[53] While his fans argued for and against him, Godfrey admitted that fame *had* left him out of touch with his fans saying, "It's no longer possible for Joe Blow to be himself with me. He becomes a phony when he's near me. . . . Where can I find friends anymore? Most people that seek me out are looking for something. [It has caused me to lose] something that has made me. I once knew all the people's attitudes. Now I'm going on memory."[54] In supporting Crosby's argument in this manner, Godfrey was able temporarily to evade the accusations that the way in which his original persona was constructed was disingenuous by blaming the dysfunctions of fame for his altered persona.

Godfrey and the Alienation of the Postwar Man

The LaRosa/Godfrey affair also reveals how a television star's construction as authentic could be undermined if his or her industrial power or business acumen was not properly narrativized. During these years, the public might question the success of an individual in the entertainment industry if they engaged in excessive behaviors or if their spending was directed away from the care for home and family. Elaine Tyler May's study of the American family in the postwar era reveals that consumption at this time reflected overall social patterns of containment and conformity. Finding that "family-centered spending reassured Americans that affluence would strengthen the American way of life,"[55] May argues that the purchasing of household appliances and other domestic accouterments (instead of luxury goods) assuaged anxieties over the decadence and immorality that could result from living *too* well. The presentation of many postwar television actors' home lives contributed to the furthering of this paradigm by presenting their labor as only a slightly more extreme version of every American's quest for a better life for their families. For example, Tinky Weisblat discusses how the extratextual representations of Ozzie and Harriet Nelson's work on television as a family business or "cottage industry" neatly "enveloped the new medium in domesticity by suggesting that the producers of television were more interested in families than in finance—and thereby helped to legitimate the TV industry."[56] A similar framing helped define Desi Arnaz and Lucille Ball's power-couple role in the industry during the same period. Their creation of Desilu and their lucrative arrangement with CBS was presented to the public as simply the happy ending to a complicated love story. Separated for years by their individual film and music careers, their work on *I Love Lucy* was often recounted as the only way they, as successful entertainers, could sustain a normal everyday marriage.[57]

In its earlier phase, Godfrey's extratextual material certainly did validate his participation in the business of television as healthy, normal, and family-centered. Furthermore, the star set himself apart from the industry's "hucksterism" by acting as a mediator of its commercial imperatives. Through his delivery style and his tempered sales pitches, he was able to act as if he were a consumer himself, trying out products and testing advertising claims. His status as a wary or cynical insider of the business increased audience identification and his claims to authenticity, for as Jack Wilson reported in *Look*, "Most people at some time would like to tell off the boss, if they dared. Godfrey did it for them, once removed, by telling off his boss."[58] Yet, with the revelation of his behind-the-scenes role as a hard-nosed program director and boss, it appeared, at least to some sections of the public, that he was much more of a company man than they ever would have expected. Rather than the kindly head of his on- and off-screen families, he seemed to have betrayed himself as a man driven by his ego and desire for power. Through the uncovering of what many after 1953 believed to be his true persona, Godfrey's image resonated with larger discourses of the effects that corporate structures had on the postwar man. The publication of books such as *The Lonely Crowd* (1950), *The Man in the Grey Flannel Suit* (1954), and *The Organization Man* (1956) analyzed and confirmed the deleterious consequences of consumerism, suburbanism, and working in corporate culture had on individuality. Yet even before the publication of such books, the ambivalence many people were feeling about the American dream of prosperity were circulating in the public sphere. David Halberstam, in his study of the 1950s, asserted, "The debate seemed to focus on the question of whether, despite the significant and dramatic increase in the standard of living for many Americans, the new white-collar life was turning into something of a trap and whether the greater material benefits it promised and delivered were being exchanged for freedom and individuality."[59] In this context, Godfrey's fall could be configured as a loss of his humanity to the alienating effect of the broadcasting business and the implied dysfunctions of fame. When Godfrey said that his success had caused him to "lose something," to disconnect from friends and the man on the street, he was speaking directly to the concerns of the populace. This particular narrative was,

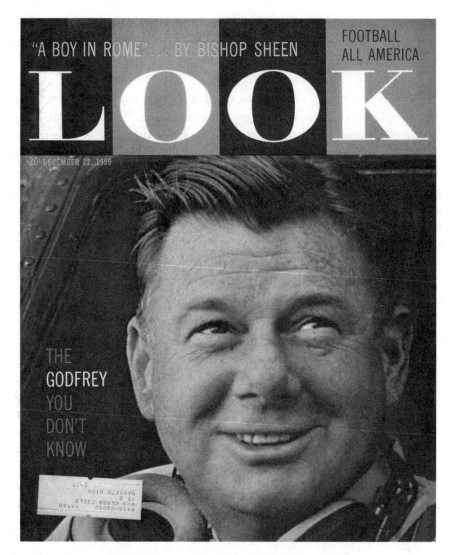

Fig. 5.6 *Look* promises to tell readers about "The Godfrey You Don't Know." 1959.

perhaps, meant to counter the accusations that Godfrey saw himself as superior to his supporting players and that he was an out-of-control narcissistic individualist, by positioning himself as simply another victim of the dark side of the American dream.

As much as Godfrey tried to blame the industry and its resulting effects for the situation with LaRosa, his career never fully recuperated from the scandal. His popularity with audiences and ratings dwindled over the next few years. A spate of firings of performers and writers of *Arthur Godfrey and His Friends* in 1955 led to more bad press for the star and seemed to further the reception of him as an uncompassionate "hatchet man."[60] Although he continued producing and starring in his three shows through 1957, when *Friends* was taken off the air because of poor ratings (ABC's *Disneyland* had come to rule that time slot), he remained on the air until 1959, when he took time off from his career to battle lung cancer. Although he returned briefly (after recovering from a surgery to remove his lung) to co-host *Candid Camera* in 1960, Godfrey soon quit television altogether.

Following his departure from show business, Arthur Godfrey disappeared rather quickly from popular memory. Although not completely forgotten by audiences of the 1940s and 1950s, many popular historical narratives that describe this period in broadcast history have neglected to include the story of Godfrey and his popularity. Academic accounts of the period also have tended to give him and his programs short shrift. It seems as though he exists as barely a footnote in the reconstruction of this era, even though his career was central to the industry's postwar delineation of personality and audience identification. The rise and fall of this particular television star speaks to the impact that the industry's positioning of television as entertainment's most "realistic" medium had on the formation of broadcast stardom. It also provides a fascinating entry into a discussion of how the signs of a star's ordinariness, sincerity, and genuineness were so tenuously located in the complex commercial environment of television.

CHAPTER 6
FOR THE LOVE OF LUCY
Packaging the Sitcom Star

By the mid-1950s, sitcom stars such as Lucille Ball, Desi Arnaz, Danny Thomas, and Ozzie and Harriet Nelson had taken hold of television viewers' imaginations as well as their purse strings. This new breed of television performer, although borrowing and recasting certain character-istics of their vaudeo predecessors, represented developing trends in the broadcast industry's management of stars and programming. In particular, shifting relationships among sponsors, networks, agencies, and indepen-dent production companies contributed to an acceleration of star-based marketing strategies that began in 1930s radio. Broadcasting stars were aggressively packaged and sold to audiences in more ways and forms than ever before, and the participation of stars in commercial processes was imperative. A 1953 *Variety* editorial noted, "The use of TV stars as pitchmen on their own shows has now reached the point where decision by some agencies on whether or not to buy a show is predicated on the willingness of the program's star to become a part-time salesman."[1]

This trend was most apparent in the proliferation of star endorsements and merchandising. For example, Ball and Arnaz lent out their names and personas for *I Love Lucy* furniture, clothing, dolls, and comic strips (among other innumerable items). Yet star images were more enduring in other ways than tie-in merchandising. With the advent of telefilm, not only could shows be broadcast nationwide in good quality but also syndication made it possible for viewers to engage in repeat viewings of their favorite programs and stars. In addition, this same technology allowed for foreign distribution, bringing international fame to television performers. Stars were also used more aggressively to sell programming to local markets. From 1953 to 1955, many agencies "traveled" their stars vaudeville style, broadcasting their programs from various local stations in front of rural or suburban audiences in order to imbricate themselves and their products within the new consumer demographic.

Ironically, just as star images were proliferating through syndication and marketing campaigns, the length and number of shows produced in a season was dwindling. Afraid of overexposing and overworking their stars as they did during vaudeo's peak, networks and agencies limited much of their fare to a smaller number of half-hour programs a season. Consequently, new comedy shows were written in half-hour segments and many already established variety shows were truncated. For example, during the 1954–1955, season *Colgate Comedy Hour* went from a one-hour program to a half-hour, *Your Show of Shows* from ninety minutes to an hour, and programs for Dean Martin and Jerry Lewis, Milton Berle, and Martha Raye were scheduled to appear only from five to twenty times a year. Noting this, *Variety* proclaimed, "The overexposed comic on TV is fast becoming extinct. The point of diminishing returns has set in and the boys who were cautioned a couple years back that they would wear out their welcome by overextending themselves on the hour formats are beating a hasty retreat into less frequent exposure and capsule editions."[2] Consequently, the half-hour format became the standard for the comedic programs, while the hour-long program was commonly reserved for dramas and specials.

Many of television's narrative and technical processes were standardized during the mid-1950s. Television historians have argued that *I Love Lucy* was *the* prototype for these standards. It has been said that the use of telefilm, the three-camera setup, the reliance on syndication for profit, the weekly production schedule, and the importance of independent packagers were all by-products of this unprecedented successful program. *I Love Lucy* was a major factor in the sitcom's ascension as the dominant format for prime-time television comedy, as the narrative trajectories were shown to be more conducive to the entertainment needs of the new consumer than the vaudeo format. Ball and Arnaz instituted many of the innovations in independent production that resulted in altered relationships among performers, producers, networks, sponsors, and audience. And, in their position as both producers and a married couple, they came to represent the newly fashioned ideal of the television star. Beginning in 1953, female comedy stars started to grow in number and in popularity as male comedians as presented in family sitcoms got tamer, more domesticated not only by their on-screen families but also by the limits of their program's mise-en-scène. What both male and female comedians had in common, however, was that, during a period of increasing commercialization in American culture, they were pressured to become more fully enmeshed in pushing the product, whether it was their own merchandising, a sponsor's product, or the strength of a television network.

Stars, Networks, and Programming

After being forced to confront the industry's talent crisis during the 1951–1952 season, the three major networks struggled to develop new talent and formats and, in an effort to appease sponsors and critics, to cap production costs. As a result, the years between 1953 and 1955 saw significant shifts in programming development, contract terms, and promotional strategies. Additionally, after witnessing the unprecedented popularity of I Love Lucy (buoyed by the lifting of the freeze and the subsequent proliferation of new stations) networks were confronted with a newly fashioned blueprint for television entertainment and stardom.

In April 1952, the Nielson Ratings Service announced that I Love Lucy had done the unthinkable: it had reached ten million households, which meant that a record twenty-three million people had tuned into its most recent broadcast.[3] This official confirmation of the program's unprecedented appeal was simply another reason for CBS and the other networks to produce more sitcoms of a similar ilk. Still believing that sitcoms were an economical alternative to star-studded and lavish variety-show productions, CBS invested much of its resources into sitcom development. With their heavy reliance on the format, CBS was the programming leader among the major networks during the peak years of I Love Lucy. Programs such as Life with Luigi, Mama, The Jackie Gleason Show (highlighting "The Honeymooners"), and Our Miss Brooks all garnered high ratings for the network from the 1951–1954 seasons. In addition, CBS still had Arthur Godfrey, the perennial ratings-grabber, as well as Red Buttons (who eventually went over to NBC in late 1953 after his ratings dropped with CBS), Red Skelton, and Jack Benny. In the first week of December 1953, ARB ratings showed that CBS had seven of network television's top ten shows.

CBS also boasted some of the lowest talent costs for these programs. The network's most expensive offering was *The Jack Benny Program*, which came in at $40,000 per half hour, while the low-cost *Toast of the Town* and *What's My Line* were estimated at $17,000 per half hour.[4] By working to cap production costs, CBS attempted to provide advertisers with top talent at reasonable prices. Even after the record setting $8-million two-year contract that the network signed with Desilu in early 1953, *Variety* heralded CBS's moves to keep costs down and contracts short: "CBS's William S. Paley has forged ahead into a new sphere of influence and affluence without involving the network in the kind of long range top-coin commitments (on talent and program patterns) that have characterized NBC deals over the past few years."[5]

CBS's major competitor was NBC. In spring 1953, *Variety* reported that NBC too had decided to end its "long range top-coin commitments," announcing that "The signposts are up at NBC. The era of fabulous TV deals are definitely over. . . . It's not a question of getting tough, but conditioning talent to the realization that the era of get 'em at any cost, so long as we got 'em deal is now over."[6] This issue became particularly relevant when NBC was forced to contend with what to do with the long-term contracted stars after their shows were taken off the air. For instance, the cancellation of *All Star Revue* in April 1953 forced NBC to keep paying Jimmy Durante even though he would no longer appear on an NBC-featured program.[7]

While showcasing one popular sitcom, *The Life of Riley*, NBC was still heavily invested in the variety format in the 1952–1953 season with *Your Show of Shows*, *The Buick Circus Hour*, *Texaco Star Theatre*, and *The Colgate Comedy Hour* still in the top twenty-five. By the 1953–1954 season, NBC's stable of top shows included a few drama series such as *Dragnet* and the *Philco Television Playhouse*, but also added new and expensive variety programs such as *All-Star Revue* and *The Bob Hope Show*. In addition, Pat Weaver's emphasis on programming spectaculars kept NBC's budget for prime-time programming at an all-time high. In September 1954, *Broadcasting, Telecasting* reported that these "extravaganza shows," which emphasized top-name Hollywood and television talent, perpetuated the "lure 'em with top stars philosophy."[8] Some of these shows did prove profitable for the network, but NBC was still far behind CBS in overall ratings and had nothing on the air that compared to CBS's *I Love Lucy*.

Falling far behind even NBC was ABC, which had only fourteen affiliates in 1951 (compared to NBC and CBS which had more than seventy a piece). ABC also did not have even one program in the top twenty-five until *Disneyland* came in at sixth place in the ratings at the end of the 1954–1955 season. As mentioned in chapter 4, the 1951 merger of ABC and United Paramount Theatres would have a positive effect on the network's ability to attract top name talent; however, this would not really begin to take effect until 1953 when Leonard Goldenson instituted his plan to find a niche for the network. According to Frank Rose, at the end of the 1951 season, "Goldenson's research showed that the talent on NBC and CBS—Jack Benny, Burns and Allen—appealed mainly to older people, not to the young families in the eighteen-to-forty-nine age bracket who were buying cars and appliances and other big-ticket items that advertisers were eager to push."[9] In an effort to counterprogram, ABC's new chief sought talent and programming that would appeal to the youth market. One of the networks only hit shows in the early 1950s was *Ozzie and Harriet*, which provided some incentive for younger audiences to tune in to watch the couple's handsome young son Ricky (who would eventually become an Elvis-like "heartthrob" for teen-aged girls). Goldenson also wanted to invest some of the thirty-seven million dollars he had pledged to put into ABC operations toward shows starring comedians George Jessel, Ray Bolger, and Danny Thomas.[10] In early March of 1953, ABC announced that it had signed Jessel to work as both a performer and a producer for the network.[11] A few weeks later and with the assistance of Abe Lastfogel of William Morris, Goldenson signed both Bolger and Thomas to their own "life in show business"–style sitcoms.

The industry recognized these deals as a significant coup for the third-ranked network. In March, *Variety* reported that ABC's new concept of signing talent such as Bolger and Thomas to "hard deals" (reasonable salaries for relatively short stretches of time) was intended to keep production costs low. The article described the ABC network's approach as one that would "make every dollar count by (1) giving the performers weekly exposure in half-hour vehicles; (2) concentrating on stars, stories, and scripts rather than on the big choruses and outsized productions; and (3) going to film in some cases, which reduces the cost to each advertiser by spreading the expenditure of the subsequent runs."[12] The publication also predicted that the acquisition of this caliber of talent would lead to the network's eventual rise: "Until now ABC has been running a poor third in the coast-to-coast web rivalry, but with its acquisition of valuable properties in recent weeks, topped by the Danny Thomas-Ray Bolger-George Jessel pacts, the 'full speed ahead' signs are posted."[13]

Both Bolger's *Where's Raymond* (renamed *The Ray Bolger Show* the following year) and Thomas's *Make Room for Daddy*, whose technical production was handled by Desilu, premiered in the fall of the following season to mixed reviews. However, according to Rose, the ratings quickly revealed that Thomas's nightclub comic character was more appealing than Bolger's, as "Television audiences were a lot more interested in watching a nightclub comic stay home with his family than in seeing a Broadway song-and-dance man try to get to the theatre on time."[14] In the long run, *Make Room for Daddy* did prove to be a solid investment for ABC as it stayed on the network until 1957 when the title was changed to *The Danny Thomas Show* and the show moved over to CBS.

Early in the 1954–1955 season, ABC initiated a million-dollar promotional campaign, including billboards, ad mats, slides, car-card drives, and promotional films featuring the stars of ABC programs.[15] More significantly for the promise of a television star system, ABC had begun working with Paramount to develop performers who would work in both the film and television media. According to *Variety*, the plan was this: "A studio singles out an unknown who shows promise. The 'new face' is brought to the ABC network and will be given the second lead in, perhaps, a dramatic airer at the first opportunity. On-camera competence and public reaction will be closely measured. One or two more TV exposures, possibly in a full-lead assignment, will be considered."[16] Goldenson announced that, after a successful run on television, the performer would be given an opportunity to star in a film at Paramount.

This type of exchange was becoming common practice by the middle of the decade. *Variety* reported in May 1955 that more television stars were going to Hollywood than ever before.[17] At that time, Warner Bros. had deals with Eve Arden, Liberace, Jack Webb, and Ed Sullivan, relationships that were said to have been inspired by MGM's success with Vincent Minnelli's *The Long, Long Trailer* (1954), a film that featured Arnaz and Ball in roles that built on their television characters. These types of deals were said to mesh "with the idea increasingly expressed that what's big for pictures is big for video, and turnabout."[18] It, of course, also was a result of the studios' increasing involvement and investment in television production. NBC not only loaned out its performers and writers to studios but also to night clubs and guest appearances on other networks in order "to give promising talent more exposure."[19]

Network strategies included developing programs in other genres besides comedy. As it had hoped, ABC was attracting younger audiences (in fact, the youngest of them all) by 1955 with shows such as *Disneyland*, *The Lone Ranger*, *The Adventures of Rin Tin Tin*, *Kukla, Fran and Ollie*, and *The Mickey Mouse Club*.[20] All networks were programming quiz shows by 1955 (many of which were hitting the top ten) and live anthology dramas, and filmed medical dramas were slowly developing significant followings.[21] Still, the sitcom would prove to be the most durable and ultimately profitable format for the medium, as viewers were infinitely more engaged with these programs' characters and domestic settings. Advertising agencies recognized this as well and their promotional and product-advertising campaigns during the early to mid-1950s centered on these elements of the format in order to stimulate identification and material imitation in the consumer-viewer.

Star Promotion and Advertising Intensifies

From the inception of standardized programming in the late 1940s, television performers were evaluated by sponsors on their ability to move product. However, during the early to mid-1950s, the strategies employed to use stars to sell programs and products became increasingly synergetic and prolific. The role that television stars played in advertising was, in fact, so essential to broadcasting's economic structure that by 1956 George Rosen indicated that "salesmanship" (along with attractiveness to housewives, and articulateness) was an essential quality in hosts of the Dave Garroway or Steve Allen ilk.[22] Ad agencies and networks used a collection of promotional tools honed in other entertainment industries to fuse the star's persona with the program and product's overall identities. These campaigns utilized approaches that combined exploitation, advertising, and merchandising to achieve a comprehensive commercial campaign. In doing so, the television industry once again acted as poacher, borrowing and recombining strategies previously employed in film, radio, and vaudeville to achieve ultimately overt yet seamless promotion that would advertise the program, the sponsors' product and television's place in the family as well as overall consumerist ideologies.

Certainly, the use of stars to promote products was not unique to broadcasting. The use of film-star images in print, and eventually broadcast, advertising for consumer products appears, at least on the surface, to be quite similar to the way television stars were used. However, film stars were not packaged quite as intensely as television stars nor were their careers so utterly dependent on their roles as product "pitchmen." Television stars were explicitly connected to a variety of products both within their program text and outside of it, while film stars were most commonly used implicitly to sell clothes, makeup, and other products placed within their films without directly addressing spectators and engaging in overt salesmanship. Indeed, film stars would appear in a limited number of ads outside of filmic texts during the 1950s, most often selling cigarettes or beauty products; yet these endeavors were not as synergistic as what was found in broadcast-related promotions. As I have discussed in previous chapters, the film industry has historically worked to simultaneously profit from yet obfuscate the affiliation between their stars and the machinations of commercialism. It was largely assumed that "rarity value" and glamour of these stars would be tainted by overt association with mass production and commodification. Nevertheless, the ties between stars and commercial products were potent and film studios worked diligently to exploit them in covert ways.

Jane Gaines has discussed how the development of the star system coincided with the acceleration of film promotion. Describing early examples of publicity and exploitation including gags, scandals, and hoaxes involving early film stars, Gaines argues that these were highly structured ploys to publicize the stars themselves, and to naturalize audience interest and its accumulation of knowledge about the star in order to get people into the theater to see the stars' films. One of the earliest examples of this was, of course, Florence Lawrence, whose death was staged in 1910 in order to court audience interest. Gaines describes this type of phenomenon in which "news" is planted to generate publicity as a way in which the film industry could simultaneously benefit from and distance itself from commercial exploits:

> The commercial motive is effectively erased as promotion disappears into everyday life, creating the illusion that nothing is being sold. To succeed, their exploitation must have a kind of transparency. A First National publicity man writing in *Moving Picture World* in 1918 thus describes this requisite transparency as a phenomenon that should "jolt the preoccupied mind of the public" while not "attract[ting] attention only to itself."[23]

In 1928, J. Walter Thompson developed its infamous Lux soap campaign, which cunningly attached the phenomenon of stardom to a household product. In order to claim that "more Lux soap was being used in Holly-wood than any other soap" or that "nine out of ten stars prefer Lux soap," the ad agency sent endless shipments of the product to Hollywood studios, thereby backing up its own truth claim. But by the 1930s, product placements and tie-ins were ubiquitous in Hollywood. The head of publicity would often work with the script, matching appropriate sponsorship with scenes that would display specific products and services. Some particularly lucrative partnerships developed out of such schemes. In 1933 MGM signed a $500,000 contract with Coca-Cola, and in 1934 Warner Bros. brokered a deal with General Electric and General Motors that allowed for the exchange of product placement for star endorsements.[24] Charles Eckert discusses this process as one that developed out of confluence of three major factors: (1) a recognition of the purchasing power of the female consumer; (2) the studios' determination to profit from product display; and (3) a star system dominated by female stars.[25] The potentially lucrative connection between female stars and spectators was most often activated through the stars' donning of particular fashions and makeup. The industry quickly believed that women would mimic what they saw on the screen, purchasing approximations of particular dresses, accessories, and cosmetics. Eckert describes this phenomenon as somewhat participatory: "Fashion worked to elicit women's participation in star and screen myth-making. Women bought star products and tested star beauty recipes, circulating ideas about star image in their own improvised 'looks'. At some point, then, it seems as though the mania for star fashions 'sprang' directly from women fans."[26] Mary Anne Doane argues that this process seems to bring the image of the star closer to the female spectator—at least psychologically—but that, ultimately, "the female subject of the consumer look in the cinematic arena becomes, through a series of mediations, the industry's own merchandising asset."[27]

With the advent of audience and consumer tracking devices, this process was refined even further in the 1940s and 1950s. More tie-ins and product placements were used in the film texts, as the film industry came to depend on market research and psychological studies on the emotional processes of consumerism. Radio during this period was able to evaluate and accommodate audience response to a particular program, product, or persona more quickly than ever before with the aid of Hooper and Crossley ratings services. In addition, radio could sustain long-term affiliations between a sponsor's product and a star persona, so that much so that it became impossible not to think of Jell-O when seeing or hearing Jack Benny's name or to

"My eyes are my trademark!"

When you see Eddie Cantor's famous banjo eyes, you look for comedy, humor, a touch of pathos—a real virtuoso performance.

And when you see a familiar brand name as you shop, you expect an equally outstanding performance — or you don't buy the product again.

That's one big advantage about living in a land where you enjoy free

choice among many fine products, each identified by its own brand name.

Leading manufacturers, seeking to win your favor for their brands, take infinite pains and a great deal of pride in bringing you wonderful products, continuously improved, representing unusual value for your money.

As you leaf through the pages of

this magazine, note how many of the products advertised here already have satisfied you. And always remember that when you name your brand, you better your brand of living!

BRAND NAMES FOUNDATION
INCORPORATED

A Non-Profit Educational Foundation
37 West 57 Street, New York 19, N.Y.

Fig. 6.1 A 1952 Brand Names Foundation ad extols the virtues of branding and uses Cantor's "banjo eyes" as an example of how a trademarked quality can provide more visibility for product.

be reminded of Rinso when thinking of Al Jolson. Clearly, some of these strategies were in use in radio during its Golden Age. What is different here is that in the early to mid-1950s emphasis on the continuing character and the domestic mise-en-scène provided by the family sitcom contributed to a richer commercial environment, one highly responsive to the industry's need for symbiotic and atmospheric promotional campaigns.

In the post-Paramount-decree era, many film stars remained trepidatious regarding excessive commercial exposure. Television stars, however, were significantly more malleable in this respect. Because of the inherent commercial obligations of television work, performers were expected from the very beginning to act as salespeople. Therefore, their careers were not subsequently hindered by multiple product endorsements. Television stars were expected to embrace their roles as salespeople and to work on the behalf of the network, sponsors, and advertising agencies. In 1954, Jessel acknowledged his acquiesce to industrial imperatives: "I, at the moment, am being sponsored by four products and have entered the television business exclusively for this year with this set idea—to sell the products that are paying for me, for I believe that is what I am supposed to do. When the sponsor asks me to personally deliver the commercial, I do it. When the sponsor tells me to take out some dialog [*sic*] or a song to make room for the commercial, I do it."[28] And in addition to the textual concerns and requirements a sponsor may have, a star often would be asked to make personal appearances or to lend his or her image to local sales initiatives. Steve Allen's image and signature were on network letters sent out to potential advertisers for his *Tonight Show*. One such letter had Allen acknowledge his role as salesperson: "Now, let me slip into my three button suit and my black knit tie so I can make with the sales talk. *Tonight* is so flexible you have to hold it with pliers to examine it . . . if you want further details, or if you already see the wisdom of joining our first sponsors, call your NBC-TV representative. He's got a family to support too."[29]

Most networks offered trade calls, mailings, advertising support, and on-air advertising, which included use of stars and their image. However, CBS would not agree to commingle advertiser's merchandising with program promotion in the mid-1950s. In a 1953 article for *Broadcasting, Telecasting* on network merchandising aids, a CBS spokesperson claimed that the network "deliberately excludes itself from [such] merchandising activity . . . having observed that such practices all too often hinder, rather than help, the elaborate merchandising plans which advertisers themselves initiate to solve their own special problems among distributors and retailers."[30] Instead, the network promised "the most effective [program] promotion in network television."[31] Yet, this went against traditional industry thinking at the time. In 1953, Hal Davis, a vice president at Kenyon & Eckhardt advertising agency, asserted, "It seems fairly obvious that a television personality should do his (or her) utmost to merchandise his sponsor's products, both nationally and locally. It takes a powerhouse

Fig. 6.2   Cantor and his "banjo eyes" are used in this S and W ad to sell juice.

of a program series to overcome non-integration of a personality and commercial."[32] Arguing that "the visible impact of television is opening up
new promotion channels never used in radio and adapting old ones which
have long been successful in Hollywood motion picture exploitation,"
Walter A. Scanlon, merchandising and promotion manager for CBS television film sales, underscored the essential role that stars play in television

promotions in general. "An actor in radio is known only by a voice, and a listener could pass him without recognizing him. In television, however, the stars are as well known as the products and brand names, and they are used to promote the programs and products in a way that was never before utilized."[33] Some of these strategies involved roadshowing television programs and talent and creating unique local promotions. Many sponsors acted as though they owned stars, requiring them to perform or appear at their request—at sales and board meetings, luncheons, plant openings and, even, their wives' charity functions. Ed Sullivan was known to be exceptionally amenable to participating in these sort of events for Lincoln-Mercury. He would not only appear wherever he went in a Lincoln but also would attend regional dealer meetings, beauty contests, and any other corporate event or publicity stunt.

In the years immediately following the lifting of the freeze, many local distributors and dealers who had invested in high-profile programming began to pressure national sponsors such as General Foods, U.S. Steel, and Colgate to travel their network programs.[34] This meant that a show's entire staff would be brought into cities such as Chicago, Boston, Cleveland, Miami, and Dallas to broadcast their program in front of a large studio audience. The broadcast would be preceded by a major print and broadcast promotional campaign (the expense of which would be shared between sponsor and local sales forces) that would often center on the program's stars. Although costly and rather short-lived, such an event could improve network relations with local affiliates, the sponsor's relationship with its local sales groups, and the local audience's interest and investment in the product, program and its stars. A 1953 article in *Variety* reported that:

> In an era when the merchandising element plays such an integral part in TV programming the distribs and dealers (who in many cases share in the cost of the programming) recognize the added values of "showcasing" their talent on a local level to hypo product sales, particularly in lagging areas and at the same time permit the talent to appraise audience reaction and values on a "Grassroots USA" level away from the major New York and Hollywood production centres.[35]

Like many of broadcasting's marketing strategies, the traveling show concept was not new. Mobilizing audience engagement on an intimate local level was the foundation of vaudeville presentation as well as for amusement parks, circuses, and early film exhibition practices. Roadshowing a particular film could generate much audience interest and "street stunts," and other forms of merchandising exploitation occurring in theater lobbies and sidewalks would attract more patrons than print and broadcast ads alone. In outlining many of the stunts publicists enacted for films in the late 1920s and 1930s, Gaines argues that these events and displays simultaneously sensationalized and valorized film products. She also notes, "In the end, what counted was that moviegoing was integrated into the rhythms of local life and that commercially produced entertainment became embedded in the existing community life."[36] This sentiment resonates with the way in which the television industry conceived traveling their television programs. Increasing audience investment in and identification with a program and its talent was an effective way to encourage regular viewing. In addition, industry insiders claimed that performers might better understand the sensibility of their audience through such local interactions. This idea in particular underscores the rhetoric that plagued vaudeo—that comedians such as Berle were out of touch with middle-class conservative tastes and the censorious atmosphere that was taking hold of television in the early to mid-1950s. Even if stars who performed their programs in front of local audiences did not actually accumulate new knowledge about their audience base, their involvement with local communities would give the public the impression that the performers were making a goodwill effort toward that end.

In an editorial written in support of roadshowing television programs, Ted Mack, host of *The Original Amateur Hour* outlined the ways in which shows and sponsors benefited from the appearance of stars in local outlets, from increased sales to improved public relations.[37] He mentioned the ways in which Ed Sullivan's appearances at Lincoln-Mercury dealer sales meetings increased sales significantly for his sponsor and how Robert Montgomery became a major touring attraction in department stores while appearing for Lucky Strike and Johnson Wax. While asserting that his own show was so successful on tour that it maintained a large waiting list of cities, Mack delineated the specific manner in which *Amateur Hour* was packaged for touring purposes:

> Dealers look upon out-of-town originations as providing the kind of publicity that money cannot buy. Our show invariably hits the front pages, perhaps for several days running. And it IS news with its many local aspirants participating, its turning of receipts over to a local cause. The leading citizens of the community sponsor our coming and we work with them for weeks to make the event a success. Thus, by the time the

show arrives, it is considered an important community project and the national sponsor, whose product we publicize each week, becomes identified as a benefactor. . . . We're not ashamed to say that we're merchandisers as well as entertainers. We may, in fact, be better entertainers because of that.[38]

One company was so impressed by the results of roadshowing television that it decided to create live programs that would remain on the road, avoiding the production centers of New York or Los Angeles altogether. In 1953, the packager, TV Roadshows, Inc. (a company created by former advertising agency television producers), developed live locally broadcast programs carrying name talent that would constantly tour seventeen major markets including Chicago, Baltimore, Boston, Buffalo, Minneapolis, Kansas City, and Indianapolis.[39] These "mobile star units," which included talent, a script girl, musical conductor, seven musicians, and a publicity person, would be financed by local companies whose ads would be integrated into the monthly half-hour "locally angled" programs.[40] The show would be regularly restaged with new plots and music, as each individual market was promised a revamped "Show of the Month" every four weeks.[41] In addition, TV Roadshows would provide a publicity promotion kit and gratis cooperative promotions such as music store cards, newspaper and radio publicity, and theater lobby displays. James Beach, the head of the TV Roadshows project, was quoted as saying "the idea is as old as vaudeville and as complex as television."[42]

A rash of sponsors traveled their programs from 1953 to 1955, but this was not the only way in which to imbricate programs into the everyday workings of a community market. Other stunts involving the program's star were often cheaper and received just as much press attention as bringing an entire production to town. A relatively early example of local cross-promotion involving local recognition of a star's persona was developed by Irving Fein, the Director of Public Relations for DeSoto. Fein proposed a stunt involving Groucho Marx look-a-likes to promote *You Bet Your Life* and DeSoto wherein the Grouchos would drive convertibles around the 170 towns nationwide that carried the program.[43] The first person in each town to say to the impersonator "you are Groucho Marx" would be awarded the car. The event would be announced in local papers and photographers and reporters would follow the Grouchos around for "extensive interviews." Fein estimated the cost of this promotion to be around $200,000. Desoto and the network expected that this local event would not only stimulate interest in the convertible but also would solidify the consumer's identification of Marx with the product brand. It also would help them get around the problem that Marx refused to participate in such stunts.

In some instances, the audiences did the traveling rather than the talent. Exploiting the female audience's interest in Sid Caesar, sponsors for *Your Show of Shows* arranged a deal with New Haven Railroad to offer half-price train tickets to 750 women to travel from Hartford, Connecticut, to New York to be in the studio audience.[44] On the train, these women received gifts from sponsors such as watchbands, Old Gold cigarettes, and copies of the *New York Journal-American* newspaper.

Program sponsors took full advantage of star endorsements both onscreen and off. Conjoining the name of the television star with the product of a program's sponsor remained of the utmost importance, yet was increasingly complicated by the mid-1950s for daytime programs as many of these programs entered into participation advertising to cut costs. Most prime-time programming, however, remained under single sponsorship by large national corporations throughout the 1950s. And, in a time when the sitcom was in favor with program packagers, the continuing character and setting of the format provided the advertiser with a more stable milieu in which to institute product identification. Indeed, instead of relying on the flexible personas of comics such as Berle, who would perform as any number of characters, advertisers were provided with lead characters in sitcoms that remained consistent in their identity and whose "lives" were set in middle-class domestic settings. This would prove to be a rich textual environment in which to attract and retain a lucrative consumer base.

Commercials were usually integrated into the text of the sitcom format. Either the program would begin and end with the program's stars interacting with the sponsor's product or the performers would use and comment on the product within the narrative of the text. Ball and Arnaz would promote Philip Morris at the opening *of I Love Lucy* first through cartoon versions of themselves (who would perform in and around a cigarette box) and then as "themselves" directly addressing the audience in their appeal for their sponsor. In commercials bracketing their programs, Burns and Allen would appear in the program's kitchen or living-room set using Carnation condensed milk in their coffee or cooking while the host Harry Von Zell (who also played a neighbor in the sitcom) would make the actual sales pitch. Furthermore, Allen or Von Zell would often extol

Fig. 6.3 Lucille Ball sells her program sponsor's product in this 1955 ad.

the virtues of Carnation inside the sitcom text while making coffee or looking up recipes. Although jarring for the contemporary viewer, these commercial spots were intended to naturalize the sponsor's product in the star/character's domestic mise-en-scène. Lynn Spigel points out that the form of these types of commercials not only "work as a graphic reminder that the story had been brought to our homes through the courtesy of the sponsor, it also served to make the advertiser's pitch appear to be in a world closer to the viewer's real life since the commercial message was conveyed by stars who came out of their roles in the story to directly address the viewer at home."[45] Yet the sponsors' products were advertised not only through distinct commercial messages but also in product place-ment. Nina Liebman notes out that the presence of the sponsor's repre-sentative on the set of sitcoms often influenced the program's script as well as its staging and construction of mise-en-scène.[46] As evidence of this, she quotes Ozzie Nelson, who recalled in his 1973 autobiography:

> One of the big advantages of being the only sponsor or of being co-sponsor was that it was possible for us, perfectly legitimately , to give them a great deal of subliminal advertising in addition to their paid commercial blurbs. For example, while we were sponsored by Hotpoint, they furnished the kitchen with all the latest Hotpoint appliances, and if we had a choice of where to play a scene we'd move it into the kitchen where Harriet could be cooking dinner or putting dishes in the dish-washer or taking clothes out of the dryer. Or even if we were eating dinner in the dining room the Hotpoint appliances could still be seen in the background. The Listerine people, of course, were not so lucky. We weren't quite ready to write in any gargling scenes—not that they didn't try to coax us (in a nice way, of course).[47]

Similarly, sitcom stars in the mid-1950s would often appear in print ads in character or on the set. Interestingly, a number of these ads also mitigated the relationship between the star and audience through the inclusion of a television camera or receiver. For example, a 1953 ad for coffee shows Burns and Allen laughing and drinking in front of a CBS camera, seemingly during a break. Two smaller photos below show a couple of television cameramen sipping coffee and a family of three watching the *Burns and Allen* program, the parents each with coffee cup in hand. The captions below instruct the reader to "Work Better! Think Better! Feel Better with Coffee!" Obviously, the ad is also promoting Burns and Allen's television program (most likely, their network contract would require outside advertisers to mention the program in relation to their stars) but the backstage feel of the ad and the literal inclusion of the family audience also works to directly encourage viewer identification and imitation. Similarly, a 1955 ad for Jackie Gleason Originals by the Manhattan Shirt Company situates the product spotlighted, next to a microphone and under a CBS camera, while Gleason looks on. A 1956

WORK BETTER! . . . Rehearsal is a busy time for George Burns and Gracie Allen, stars of the *Burns and Allen Show* on CBS Television. But they always "take five" for a "Coffee-break"! Coffee's pleasant lift makes hard work seem lighter, helps to ease the strain. The best break in anybody's working day . . . is a "Coffee-break"!

—give yourself a "Coffee-break"!

THINK BETTER! . . . TV cameramen have problems on the *Burns and Allen Show*. And they solve them over coffee! Coffee gently stimulates your mind, helps you stay alert. A delicious aid to clear thinking . . . is a "Coffee-break"!

FEEL BETTER! . . . The whole family enjoys *Burns and Allen* . . . with a cheerful cup of full-strength coffee! At mealtimes or in-between — at home, at work, or in your favorite restaurant — give yourself a "Coffee-break"!

Coffee always gives you a break!

PAN-AMERICAN COFFEE BUREAU, 120 Wall St., New York 5 · Brazil · Colombia · Costa Rica · Cuba · Dominican Republic · Ecuador · El Salvador · Guatemala · Honduras · Mexico · Venezuela

Fig. 6.4   Behind the scenes with Burns and Allen.

ad for Aunt Jemima pancakes, although it does not picture a television camera, includes individual shots of the Nelson family seemingly in character and on the kitchen set of the *Ozzie and Harriet* (1952–1966, ABC). Each character makes a type of pancake that best suits his or her personality:

Ricky likes 'em tiny so he can eat a hundred.
Harriet makes 'em thin and dainty.
David's are huge—like his appetite.
Ozzie flips 'em—and sometimes misses.
All the Nelsons like Aunt Jemima's flavor best!

These print ads underscore what advertisers were trying to achieve in their broadcast commercials—to mesh the personality of the product with the star/character's image. Motivation research (such as the work done by Ernest Dichter) was an essential tool for advertisers and broadcasters during this period. Most of the results of this psychologically based research emphasized the effectiveness of associating a product with a personable television personality. The sitcom star representing contemporary consumerist and familial identities was often an ideal, living, interactive billboard on which to place product associations. When considering the processes whereby advertisers and networks mobilized these ideals through their programs and characters, it's important to remember the social historian Elaine Tyler May's claim, "Consumerism in the postwar years went far beyond the mere purchases of goods and services. It included important cultural values, demonstrated success and social mobility, and defined lifestyles."[48]

An important element added to the televisual representation of consumerism during the early to mid-1950s was the female television star. As an effective stimulator of material imitation in the lucrative female audience, women starring in comedic roles, particularly those in sitcoms, became effective guides for 1950s consumerist and domestic ideologies.

Introducing the Domestic Feminine

Although daytime programming was arguably the most effective way for the television industry to reach the postwar homemaker, prime-time programming also had to be designed to appeal to this prime demographic. One obvious way to make such a connection would be to emphasize women for identification purposes. Yet female comedians historically had a contentious relationship with broadcasting. Thought to be generally unappealing, if they embodied the brash vaudeo characteristics or acted as a program's prime host or announcer, women comedians were a rare commodity in very early television. But, by the time the sitcom was on the rise, their cultural potency and commercial competence had finally been recognized.

Fig. 6.5   The Nelsons express their individuality through pancakes.

A few female comics such as Martha Raye were able to survive as hosts in variety formats, but most women who made it into television during the late 1940s and very early 1950s partnered with a male star in a comedy team. Although their target audience was usually predominately female, producers and network heads were reluctant to place a single female comic at the center of a program unless the show was a sitcom. This was true for

Fig. 6.6 Martha Raye was often described as a grotesque of sorts, due, in part, to her "Mammoth-Cave mouth." Library of American Broadcasting.

radio as well, as commentators and industry insiders believed that a female voice was not as well suited for work as an announcer as a male voice. In their 1951 book on radio and television advertising, journalism professor E. F. Seehafer and advertising executive J. W. Laemmar reasoned that the innate cadence or pitch of a women's voices was annoying to listeners for both technological and psychological reasons:

> Ever since the earliest days of radio-telephony, men have generally been used as announcers. Their voices reproduced far better over the first radio sets than women's voices. As engineering standards improved it became possible to reproduce women's voices more truly, and many stations experimented with women in announcing roles. However, listeners seem to prefer the male voice, which is richer in quality and in overtones and deeper in pitch. Today, when used at all, women announcers are employed on programs aimed at the feminine audience. In very few cases are they used for general announcing roles. One of the biggest objections to women announcers is their tendency to sound affected or "sugary," qualities which irritate radio listeners.[49]

Not only were qualities of voices at stake; so, too, were other prejudices. The aforementioned Weiss & Geller team of social scientists agreed that female audiences have a great suspicion of "women who know too much." Therefore, the "[d]ictatorial certainty and the air of perfection" of female television announcers "are reminders of a nagging mother telling daughter exactly what to do and exactly how to do it. We all know how daughters react to nagging mothers."[50]

Ernest Havemann, in an article on Imogene Coca for *Life*, asserted that the reason for the scarcity of female comics, rather than announcers, had to do a different type of inherent unattractiveness:

> The good comedienne, of course, is a rarity, especially in America. Most American women, particularly the determined kind who are likely to go into the theater, like to be glamorous. Few actresses go in for comedy unless they have some physical peculiarity which forces them against their will to abandon romantic roles. Martha Raye, for example, has a Mammoth-Cave mouth which no make-up can disguise. Cass Daley has buck teeth. Since they drift into the business by necessity, comediennes as a rule seem to view their jobs with distaste or even with humiliation, and they go about their work with a vicious energy that suggests they are hell-bent on destroying their music, scripts, stages, microphones, audiences and possibly even themselves. The theater world regards them as a kind of necessary evil, as likely to set the customer's teeth on edge as to amuse him. [51]

Despite such harsh and misogynistic pronouncements of all female comic as grotesques, quite a few women did manage to succeed in the field in television's early years. Even before the unprecedented popularity of Lucille Ball in the mid-1950s, women such as Raye, Allen, Coca, Gertrude Berg, Audrey Meadows, and Joan Davis helped achieve and sustain strong ratings for their television programs. They did this in part like many of the successful male vaudeo stars. Denise Mann demonstrates how Raye, like Benny, often positioned herself as a fan of the Hollywood stars that would appear on her program. Thus, Raye combined the intimacy of television through the female audience's identification with her as "ordinary" within the spectacle of Hollywood glamour.[52] Attempting to capitalize on this negotiative strategy the sponsor, Hazel Bishop, had Raye demonstrate its cosmetics in testimonial form at the beginning of the program. She would mock more traditional cosmetics ads in an over-the-top style by mugging outrageously for the camera. Raymond Spector, head of Hazel Bishop, Inc., complained to *Sponsor* about Raye's performance as a product salesperson as he announced the cancellation of *The Martha Raye Show* in 1955.[53] The magazine reported that "[Spector] is not interested in Martha Raye's high ratings but the question of 'intensity of viewing . . . whether the loyalty and affection which a star generates flows over to the commercials.'"[54]

Even though Raye attracted a significant viewing audience, Spector believed that she was unable to project properly the glamour, sincerity, and sentimentality he thought was needed to move cosmetics.

Spector's comment is fairly representative of the belief that female comedians lacked certain essential feminine characteristics and that female television comedians were, in many ways, much more akin to their viewers than Hollywood stars were. Still, female comedians were spectacular in their abilities. Their outrageous physical antics, adroit verbal play, and character portrayals all made them seem quite different from an "ordinary" woman. In 1953, Jack Gould claimed that television had proven itself to be a more hospitable environment for female comics than any other entertainment form:

> Women comics, however brilliant, have always been outnumbered by the male of the species on the stage and screen. One of television's accomplishments has been to bring the distaff clowns into virtually equal prominence with the males. The rise of the comedienne in TV may be attributed to the nature of the medium itself. Since the TV audience is the family at home, the domestic comedy, revolving about the woman of the house, is a natural formula.[55]

A female cast member (as long as she wasn't a "sweater girl" or a grotesque) had already lent respectability to the vaudeo aesthetic, particularly when she mobilized signifiers of the domestic. In exchange, the codes of femininity were often reasserted onto the female comic figure, keeping the male masculine in comparison. In the majority of vaudeo "boy-girl" acts, the woman commonly portrayed a wife, romantic interest, or an assistant. As was the case with Raye, a liminal gender identity could prove problematic in the context of 1950s commercial television. But as a part of a "boy-girl" act, a female comic's femininity was asserted even if she played with its tropes. Moreover, by performing within a domestic context (either textual or extratextual), a female comedian's transgressive qualities could be tempered or contained. In describing Coca, Caesar's partner on *Your Show of Shows*, Havemann favorably compared her to other female comics: "Miss Coca, a good comedienne and a non-destructive one, has no serious physical defects. Her figure is adequate and her face, while not beautiful, would hardly stop a clock. She has no trouble finding a husband who still compliments her regularly after 16 years of marriage."[56] Although Coca could distort herself into grotesque female characters, her femininity was never in question. In part, this was because of her portrayal as a demure, married woman in "real life." But, perhaps more significantly, it was also a result of her coupling with Caesar. Time and again, profiles of Coca told of how television viewers believed Coca and Caesar were actually married in real life. In an article on Coca, a writer for *American Weekly* suggested that:

> Every Saturday night, an estimated 25,000,000 Americans turn off their television sets at the end of the giant NBC-TV program "Your Show of Shows," and sit back, still laughing, to recall Comedienne Imogene Coca's antics with Comic Sid Caesar. Then several million of them remark to one another, "She and Sid certainly are a terrific husband and wife team." Several million others add, "She sure must be a clown around the house!"[57]

In a 1951 interview with *Cosmopolitan*, Caesar provided a personal example of such a misconception:

> [Caesar] thinks the best evidence of the impression they have made upon the public as a team occurred when he took his wife, a shapely blonde, to a night club. They had a drink, danced, ate dinner and prepared to go home. As they were leaving, Caesar overheard an elderly woman say: "Isn't that terrible? Just like an actor, traipsing [sic] around with a long-legged showgirl and leaving his sweet little wife, Imogene, sitting home tending the baby. Poor thing."[58]

The viewer's assumptions of a marriage that never existed probably resulted, at least in part, from her frequent exposure to other famous broadcasting husband-and-wife teams such as Burns and Allen, Fred Allen and Portland Hoffa, Benny and Mary Livingston, and Ozzie and Harriet Nelson. Still, readings of the comedy duo as married in the light of innumerable articles on Caesar and Coca's respective real-life marriages is intriguing. Perhaps this suggests that audiences harbored resilient assumptions about the male/female teams on television and their inextricable partnership with one another.

For Gracie Allen, the difference she brought to her coupling with Burns was not only located along the lines of gender. Allen's Irish Catholicism was frequently emphasized in relation to Burn's Jewish heritage. A reporter for *Liberty Magazine* who covered Burns and Allen's early radio career inquired directly into Allen's background:

I had heard various conflicting reports about Gracie's nationality, so I questioned her about it. "I'm Irish," she answered. "You didn't think I was Jewish, did you?" "Yes," I said. "I knew that George was and I heard that you were, too." "I'm Irish as Paddy's pig," Gracie laughed, "and proud of it."[59]

In press reports appearing in the early to mid-1950s, Allen was often asked how she coped with her mixed marriage. Her response was usually that the religious differences had no impact on their marriage. For example, Allen told the *Woman's Home Companion* in 1953 that, "There has never been any religious problem. George never suggested my altering my faith and if I wanted to go to church five times a day that would be all right with him—so long as I was at the studio on time for the show. It was George who insisted that Sandra and Ronnie be brought up as Catholics, although I must say that the only one in our house who likes fish on Friday is George."[60]

Allen started her career in vaudeville as one of the "Colleens" in an Irish act. She initially planned on being a dancer, but found that she "was a natural" at comedy.[61] After pairing with Burns in the 1920s as a not-so-popular vaudeville act (they were regularly used as a "disappointment act"—an on-call position for cancellations), Allen came into the role that would later make her famous on radio and television. Burns wrote the basics of the character for her as well as much of her dialogue. Although some considered "Gracie" to be just another "dumb Dora" act, Allen found her to be a challenging and unique character. "Now there had been many dumb Doras on the stage but the Gracie character was different," said Allen. "The other dumb Doras wore funny clothes and said funny things. . . . I wasn't funny, I was wide-eyed and I never fell down or went slapstick or made funny faces and I wore the prettiest, smartest clothes I could buy. The character George developed for me is actually a sort of caricature of me and is based on what I'd call illogical logic. Every gag makes sense."[62] Patricia Mellencamp concurs with Allen's view of her character, writing:

Derailing the laws of syntax of language and logic, [Gracie's] technique was a referral back to either the nearest or the most unexpected referent as a comic turn on the arbitrary and conventional authority of speech (and she would continually break her own rules just when her friends and we caught on). She baffled all of the male and most of the female characters, concocting improbable stories and schemes that were invariably true in amazing circumlocutions which became that week's 'plot.'[63]

When in character, much of Gracie's difference was expressed through linguistic play, although, as Mellencamp suggests, her difference was not just expressed in opposition to her husband or other male characters but from other women on the show as well. Still, the female characters, although often frustrated by her confusing ways, were more accepting and understanding of her alternative logic.

The Burns and Allen Show was a sitcom/variety format hybrid. Burns would employ direct address in monologues and asides, as he was the omniscient participant in and narrator of the sitcom plots and a host of the musical segments. Burns always played the straight man for the antics of his on- and off-screen wife. As a continuation of the real-life pairing of Burns and Allen that was exposed and explored throughout their radio careers, the team entered into television once again using their real names and their real children and deliberately obscuring the difference between their characters and their "authentic" identities. Although their ethnicities and/or religion were never directly addressed on air, their standing as a couple in a "mixed marriage" was widely circulated through extratextual materials. If their backgrounds were ever confused, however, it was Allen who was perceived to be Jewish, not Burns as Irish. Burns's Jewishness was the dominant ethnic identity in this couple, in large part simply because of his standing as an ex-vaudevillian comic who befriended and was of similar professional standing with Jack Benny, Eddie Cantor, Al Jolson, and Milton Berle. Indeed, as discussed earlier in this book, the male vaudeo star was assumed to be ethnic, and most often Jewish.

In the years preceding the professional rise of Ball and Arnaz, Burns and Allen were perhaps the most domesticated comedy pair on television. Not only did their real-life marriage and suburban lifestyle present a picture of domestic bliss but also their on-screen personas reinforced particular gender roles of heterosexual coupling. Their program's sitcom plots—which played off of assumptions about their real-life marriage—were set in middle-class suburbia, and the products that they sold on the program, such as Carnation processed milk, were geared toward the female housewife consumer. As part of a married couple, Burns did not exude the destabilized masculinity of fellow vaudeo performers Benny and Berle. In fact, Burns and Allen were the precursors of domestic sitcom performers such as the Nelsons, who dominated television in the middle to late 1950s.

Gertrude Berg, in her role as Molly Goldberg in *The Goldbergs*, had been one of radio's first female stars whose fame did not depend on her

Fig. 6.7   Gertrude Berg as Molly. Library of American Broadcasting.

coupling with a male partner. Her television program of the same name first appeared on the air in 1949, with Berg acting as star, head writer, and producer. Focusing primarily on her immigrant character's struggle to foster a typical American family in the Bronx, it was said that the Goldberg family was a realistic representation of Jewish-American urban life. The novelist Charles Angoff wrote in 1951 that the Goldbergs embodied the "Neurotic tension, despair, ecstasy, conniving, kindliness, back-biting—in short, the normal life of Bronx and Brooklyn and Manhattan and Chicago and Boston and Philadelphia and San Francisco Jews." Angoff went on to argue that "Molly Goldberg, indeed, is so basically true a character that I sometimes think she may become an enduring name in the national literature. She is the prototype of the Jewish mother during the past twenty-five years."[64] Critics and audiences alike heralded the program for its ability to deal with issues pertinent to the Jewish immigrant lifestyle—particularly in relation to the often bumpy road to assimilation that many such families experienced. Berg, however, was quick to point out that her character and program's narrative trajectories were not limited to the experience of Jews, but were in fact universal. In an interview with *Commentary*, Berg said that she would not include "anything that will bother people . . . unions, politics, fundraising, Zionism, socialism, intergroup relations" in her scripts. Her reason was that "the Goldbergs are not defensive about their Jewishness, or especially aware of it. I keep things average. I don't want to lose friends."[65] The commingling of the specifics of the New York immigrant experience with the focus on "average" American everyday domestic life made Molly Goldberg an interesting and long-lasting character. Berg's ethnicity was tempered by her comedic abilities and adherence to traditional family values.[66] Joyce Antler contends that, "Berg's brilliance was to wed the iron qualities of traditional East European Jewish women with a charm and humor that counteracted the threatening aspect of their power."[67] The Goldbergs' eventual move from the Bronx to the suburb of Haverville, along with the program's adoption of the half-hour format by 1953, furthered the characters' assimilation and provided an even more hospitable environment for advertising.

Although comedians such as Berg and Allen did much to pave the way for the female television star, it was Ball's unprecedented popularity that really signaled the potential power of the women comedians in television. After many years of working in Hollywood as the self-professed "Queen of the B's," Ball found her most significant cultural impact in the television sitcom format. The actress had explored her comedic abilities in her film work, and, as Alexander Doty points out, "What made Ball different from most previous American women film comedians was her combination of slapstick with both sexuality and domesticity."[68] Yet much of Ball's sexuality had to be tempered and her image deglamourized in order to better suit the domestic and commercial setting of television. Nevertheless, Ball's negotiation of traces of her construction as Hollywood star with her sitcom housewife characterization did present intriguing ruptures in *I Love Lucy*'s representation of feminine domestic containment, but it also helped construct an environment for advertisers that could sustain the appeal of a beautiful star who could be easily read and identified with as a "real" American woman. Ball's image, performance style, and business practices would substantially alter the production of stardom in television as well to assert the dominance of the domestic feminine in television programming for many years.

Starring the Sitcom

In 1953 *Variety* announced, "A new breed of comedian, shaped by the pressing needs of television, is in the ascendance." Arguing that television needs more than just a stand-up routine at this point in its history, the magazine described the industry's move away from the "Milton Berle method" to a marked preference for the situation comedy. The article continued: "It's pointed out, however, that situation comedy provides for only one characterization rather than many and that for a new comedian looking forward to a long tele life the comedy repertory is a better pitch. Situation comedy tends to put all the yocks into one script basket, whereas repertory provides diversity."[69] In other words, some doubt existed, at least in this instance, about whether the audience could tolerate long-term investment in sitcom characters on television. By the following year, *Variety* acknowledged that the networks were focusing less on particular stars and placing more importance on the quality of the production as a whole, stating that "a comic is only as good as his material." [70] In announcing the "Death of the Television Star," the author of the article wrote: "Today it's conceded that even the best of the star crop is only as good as the vehicle; that the 'star system' as the film industry knows it, hasn't got a chance in a medium that uses up personalities at such a clip."[71]

Nevertheless, during the time that this debate was going on in the trades, Ball's star was rising to previously unforeseen heights. Hal Humphrey pointed out in one of his columns for the *Mirror* that Ball's success had surmounted industry expectation: "When 'Lucy' hit TV over a year ago a well known movie producer here said of Lucille Ball, 'She'll kill herself with the public in six months. No entertainer can come into a person's home every week and not wear out his welcome.' If there is an explanation to why Lucille still is on TV and riding the crest of all the viewing surveys, it must be that many viewers do not look upon the gal as an entertainer."[72] Instead, Humphrey believed that the audience thought of Lucy as their neighbor and were therefore highly engaged with every small occurrence that happened in her on- and off-screen life. Harry Ackerman, the head of network programs at CBS in 1953, took a stance that recognized the significance of both the performer and his/her material. Citing the television audience's attachment to the character of Lucy, Ackerman wrote, "With such evidence of the importance of individualities in relation to the vehicle, it becomes more obvious every day that 'the play,' alone, is NOT entirely 'the thing' in the birth of a new situation comedy series for television."[73]

The sitcom format did take much of the attention and workload off of the sole host or star of a comedy program; however, rather than diminishing the importance of stardom in television, the format produced its own type of star—one with an on- and off-screen persona deeply enmeshed in the program's character and mise-en-scène. Thus, the writers who constructed the character as well as the supporting cast became even more influential in the construction of the star's persona. The setting, too, was an essential factor in character development, representing the character's socioeconomic standing as well as taste, aspirations, and relationship to family members and friends. In an age when it was largely understood by the advertising industry (as well as the general public) that an individual's personality could be inferred by the products, home, and appliances she purchased, I am assuming that the television audience would be highly attuned to the significance of such household fixtures on the sitcom set.[74] Whether this, in fact, occurred, the sponsors of such shows were depending on this supposition and made decisions based on it.

The family sitcoms popular in the early to mid-1950s, such as *I Love Lucy, I Married Joan, Burns and Allen,* and *Make Room for Daddy,* retained many traits of their vaudeo predecessors. These programs continued to incorporate theatricality as an essential element in their construction of space and narrative. The most obvious strategies included the staging and framing in proscenium style, shooting in front of an audience in stage-play style in continuity, the continual referencing of life in show business, and the inclusion of musical and comedic performance into the text.[75] Furthering the reflexivity of these programs, most of the lead characters were given jobs in entertainment—usually mimicking the stars' own career trajectories. Thomas played nightclub singer and comedian Danny Williams, Burns and Allen played themselves, Ray Bolger portrayed a Broadway star living in suburbia, Arnaz was bandleader Ricky Ricardo, and Ball's Lucy's nurtured her desire to be in show business despite continual frustration. Even Joan Davis's character Joan Stevens gave up a career in show business in order to become a housewife. Spigel contends that the theatricality of this genre spoke to how suburban middle-class family life in the 1950s was itself artificial, as social convention required a type of performance of gender, class, and ethnicity. She writes, "Given the emphasis on social performance and spectacle display in postwar culture, it seems reasonable to assume that the genre of family comedy, with its self-conscious reflections on the theatricality of everyday life, might well have struck a familiar chord with audiences at the time."[76]

The fabrications of domesticity also were issues for the male lead performers of family sitcoms. As on-screen fathers and husbands, men such as Arnaz, Thomas, and Davis portrayed softer versions of their former nightclub and stage personas. As Ricardo, Arnaz was often outsmarted by his wife or teased for his inability to properly master the English language, thereby failing to completely contain or control the happenings in his home. Even in extratextual material, Arnaz was often depicted as a sensitive, and sentimental family man, who would like nothing more than to spend his nights at home with his wife, children, and dogs (which stood in stark contrast to the rumors of his unrelenting womanizing). Although they were not really feminized by their characterizations, the husbands and fathers represented in the early 1950s family sitcoms often were fallible, bumbling, or in some manner ineffectual. Even Ozzie Nelson was limited in his ability to achieve what was expected from men in the 1950s, as he was never seen working at a job to provide for his family. In fact, audiences were never told what Ozzie did for a living and the character spent most of time in the yard or in his living room. When a career was highlighted for a male character it usually was in entertainment. This strategy allowed for the family sitcom to act as a backstage component to the lives of men in show business.

Family sitcoms of this period also borrowed elements of the slapstick physical performance style perfected in vaudeo and continued the tradition of broadcasting in front of a live studio audience, and, in the case of *Burns and Allen*, direct address. Ball's rubbery pratfalls, zany movements, and repertoire of elastic facial expressions were the comedian's most spoken about and beloved characteristics. A few months after the series' debut, one *Newsweek* critic compared Ball to one of television's preeminent vaudeo hosts of the 1951–1952 season: "The clownish costumes, mugging and pratfalls the comedienne indulges in have earned her the sobriquet of the female 'Red Skelton.'"[77] The *Hollywood Reporter's* review of the show's premiere also compared Ball to Skelton as well as other physical comedians:[78] "She combines the facial mobility of Red Skelton, the innate pixie quality of Harpo Marx and the daffily jointless abandon of

Fig. 6.8 Ozzie and Harriet. Library of American Broadcasting.

the Patchwork Girl of Oz, all rolled into one. She is a consummate artist, born for television."[79]

Although her physical comedic prowess rivaled that of any early vaudeo or vaudeville performer, the assimilation of this element into a domestic setting and narrative not only eliminated much of the broad and often risqué humor found in early variety shows, but it also toned down Ball's image of beauty, glamour, and sexuality, which she had so assiduously honed during her film career. This contributed to Ball's reception as authentic and family-friendly by audiences and critics. Jack Gould, in an article that tried to answer the question of why millions of viewers loved "Lucy," asserted that the audience looked beyond her spectacular comic antics to find a common housewife:

> Unquestionably, she is the unrivaled top TV comedienne of today, a complete personality blessed with a very real and genuine comic artistry. . . . But the most durable and recognizable quality conveyed by Miss Ball—perhaps it is the real heart of "I Love Lucy"—is wifely patience. Whatever the provocation or her exasperation, she is always the regular gal and wonderful sport. On stage and off, Miss Ball is a person.[80]

Yet not all commentators were pleased with the obfuscation of the former glamour girl's exceptional beauty. *Life* complained that "her fine face has been battered with pies; her stunning figure has been obscured by baggy-pants costumes; her adeptness at comedy has been translated into slapstick pratfalls, and her versatility almost completely ignored."[81]

Indeed, the placement of the vaudeo-inspired elements of the family sitcoms of this period into a domestic setting with continuing characters and episodic narratives modified not only the structure of television comedy but also the way in which the performers were received. The reflexivity assumed in vaudeo was intensified in the family sitcom as these programs overtly promised to represent their stars' authentic off-screen relationships and daily lives. This became most apparent in the massive following that Ball and Arnaz garnered between 1952 and 1956. "The trouble with Lucy is that her real life is so much like her reel life," a friend of hers told a reporter for *Look* in June of 1952—a statement that supported the article's thesis that, "[t]here has never been a divorce between the public and private lives of the Arnazes, and their TV comedy is no exception."[82] The primary event which demonstrated the television audience's complicity in Ball and Arnaz's blurring of their on- and off-screen lives was the commingling of the birth of Desi Arnaz Jr. with that of the character "little Ricky." After battling the Biow Agency and Philip Morris to showcase Ball's pregnancy on the show and then managing the morally trepidatous subject by calling in a priest, Protestant minister, and rabbi to approve "enceinte" episodes, the executive producer Jess Oppenheimer, Ball, and Arnaz arranged to schedule "little Ricky's" birth to occur the same night of Ball's scheduled cesarean section. On January 19, 1953, forty-four million (a 71.1 ratings share) viewers chose to watch Lucy's television delivery rather than see Dwight D. Eisenhower's inauguration. In a *Look* article recalling the public fascination with the event, Oppenheimer remarked,

"That's the magic of Lucy. The things that happen to the Ricardos happen to everyone in the audience. We call it 'holding up the mirror.'"[83] By "holding up the mirror" to their stars' domestic life, the creators of *I Love Lucy* also attempted to cultivate a close identification between the female consumer and the program's stars/characters. In addition, they encouraged such an identification to extend into imitation through the consumption of Lucy's image and the show's entire mise-en-scène.

Material imitation was based in the viewers' ability to identify with the stars textually, and involved the production and purchase of consumer goods that the stars/characters supposedly used or were directly associated with. Obviously, the sponsors of these programs wished to cultivate this in order to sell Carnation milk and Goodyear tires on *Burns and Allen* and Philip Morris cigarettes on *Lucy;* yet other manufacturers extended the domestic world of their stars by offering audiences the opportunity to acquire bits and pieces of the mise-en-scène, such as the Hotpoint appliances in the Nelson household. Interestingly, in the case of *Lucy,* the sponsor was not as successful at making this connection as was the merchandiser. Philip Morris, who threatened to drop Lucy at the peak of the show's popularity, claimed that the show did not sell enough cigarettes to make its investment worth it. Yet, the vast amount of merchandise that came out of this program sold at unprecedented rates.

According to Bart Andrews, thirty-two thousand Lucy aprons and eighty-five thousand Lucy dolls were purchased in a thirty-day period in 1952.[84] The following year, highly sought-after Little Ricky dolls, nursery furniture, chair and desk sets, and *I Love Lucy* bedroom suites (one million of which were said to have sold within the first ninety days) appeared in stores.[85] In a fascinating recapitulation of the viewer's experience of watching *Lucy,* three-dimensional picture magazines of the set and characters (which came equipped with Polaroid eyeglasses for a 3-D effect) provided fans with the intensely voyeuristic pleasure of lingering with the program's textual space. Fans also could express their identification with the show through the donning of outfits made to capture the characters of both the on- and off-screen characters of Lucy and Desi. "Lucille Ball"–labeled sweaters, blouses, dresses, lingerie and jewelry were sold alongside racks of Desi smoking jackets, robes, and something called "Desi Denims" (which neither the character nor Arnaz were ever really known to wear). In an interesting turnaround of the merchandiser forcing clothes on the characters, Arnaz and Ball were contractually required to wear matching pajamas on the show to increase sales in department stores

Fig. 6.9   GE uses the popularity of *I Love Lucy* to sell television sets.

and shops. The merchandising of *Lucy* extended to other media as well. Besides the recordings of episodes that were played on the radio, Columbia released two hit *Lucy* singles: the program's theme song and "There's a Brand New Baby in Our House" sung by Arnaz and released to accompany the birth of "little Ricky." In addition, Dell put thirty-five *Lucy* comic books on the market between 1954 and 1962.[86]

Although Lucy is an extreme example of the type of program merchandising, it was nevertheless representative of the centrality of the mise-en-scène in the construction of sitcom star images. Stars were replicated through consumer products bearing their names and images and their living room settings were used as salesrooms for a particular middle-class domestic lifestyle. As mentioned earlier in this chapter, in *Burns and Allen* the living room and kitchen sets were used in the commercials for Carnation that bracketed the program. *Burns and Allen* were usually present in these commercials, silent and drinking their coffee or cooking with the product, while Von Zell made his sales pitch. On *The Goldbergs*, Molly would discuss the health benefits of her sponsor's product Rybutal vitamins while leaning out her window as if chatting with a neighbor. *Broadcasting, Telecasting* heralded Berg's salesmanship asserting that "[a]long with her strong audience appeal, Mrs. Berg is undoubtedly a sponsor's delight too. She is handling the commercials on the show in an ingratiating friend-to-friend approach to selling that is sure to boost vitamin pill sales wherever the show is carried."[87] The use of the settings in sitcom commercials helped reinforce the material relationship between the program's stars/characters and their audiences by asserting the domestic mise-en-scène as a space in which consumerism and characters intermingle. The sponsors hoped that the coopting of the program's setting would activate the female consumers' material imitation of their favorite program and star. But some settings were believed to be more fruitful than others. When the Goldbergs' assimilation into the American middle-class sitcom brought them to the suburbs, their new mise-en-scène was sold to advertisers as a more fruitful environment for products. A 1956 ad for the program announced:

> For audience impact . . . salesmanship . . . and top product identification you can't beat . . . Molly.
> Thirty-nine new half-hour shows.
> New Story Line: warmer and more entertaining than ever before.
> New settings: Molly's new home in the heart of Suburban America.
> New situations: with Molly making friends with all her new small town neighbors.[88]

By imbedding the sitcom star within the middle-class domestic lifestyle, these programs managed to intensify their claims of representing "reality." As Mary Beth Haralovich contends in her analysis of the sitcoms of the late 1950s, viewers responded to the intricate stage sets and suburban content of programs such as *Leave it to Beaver* and *Father Knows Best* by praising them for their realistic representations of middle-class life.[89] Despite social conditions that spoke to the contrary, critics and viewers perceived the interactions among the programs characters as accurately depicting middle-class family cohesion and morals. Although the early sitcoms addressed in this chapter had not yet reached the level of verisimilitude in its mise-en-scène nor dealt with small family issues and foibles, shows such as *I Love Lucy, Burns and Allen,* and *Make Room for Daddy* were a significant step in that direction, acting as mediators for the exiting of the stagy variety format and the introduction of the suburban domestic sitcom of the late 1950s.

Filmed sitcoms such as *Lucy* offered their audiences additional opportunities to explore mise-en-scène and character identification through repeated viewings. Live programming offered audiences only a single chance to experience a specific program episode and, for those on the West Coast, provided a fairly rough kinescoped vision of that episode. With film, however, repeat viewings were possible, profitable, and of good quality.[90] Repeats enabled the viewer to either watch an episode he/she had missed the first time around or to watch an episode at least one more time. This had the effect of increasing audience familiarity with a program—enhancing its knowledge of the plots, characters, stars, and settings—as well as upping its investment in the viewing process. Through repeats, certain television stars became a constant presence in the viewer's home—like an ingratiating neighbor with whom you are extremely familiar.

It is possible that Lucy's reception as an average housewife or recognizable representation of a viewer's neighbor was the reason behind her successful avoidance of the blacklist in 1953. In September of that year, Ball was accused by William Wheeler, an investigator for the House Un-American Activities Committee (HUAC), of being a member of the Communist Party. Although the actress admitted to registering as a member of the party in the 1936 primaries, she claimed that she and other members of her family only did so as a favor for her ailing leftist grandfather. Stating that her maternal grandfather was "the only father I knew," Ball explained to reporters immediately after the news broke nationally that "the reason we did it was because my grandfather wanted us to. At that time, it didn't seem at all wrong to try to please him."[91] Her framing of the issue as a family matter seemed an effective strategy in light of her already established persona as a dedicated wife and mother. Even the potentially deleterious matter of her husband's ethnicity was subverted during her meetings with the press as Arnaz took on the issue directly. "I was kicked out of Cuba because of communism. We despise everything about it," Arnaz asserted. "Lucy is as American as Barney Baruch and Ike Eisenhower."[92]

The trial in the press of Ball was somewhat of an aberration in the usual manifestations of the Red Scare in the industry. Not only did Ball come out of the ordeal relatively unscathed, but the support that she received from viewers spoke to the power that the actress had with her fans and marked one of the first public outcries against the communist witch hunts of the 1950s. Besides the Ball case, 1953 marked the linking of an enclave of the American Federation of Television and Radio Artists (AFTRA) with anticommunist advertisers and individuals in the American Legion to form Aware, Inc., which virulently attacked the supposed communist influence in broadcasting through the establishment of another blacklisting clearinghouse.[94] It wasn't until 1955 that other members of AFTRA managed to end the relationship of their union's anticommunist minority with Aware.

It was in this climate that Ball's fans signaled their support. Immediately after news of the charges was released thousands of letters and telegrams were sent to Ball, CBS, and Philip Morris in support of the actress. After waiting briefly to gauge the public's response to the situation, the sponsor and network came out publicly in support of Ball. The *Los Angeles Daily News* reported on September 14, 1953, that "Lucille Ball, her husband and her sponsors sighed with relief yesterday, convinced the voice of the public could be heard across the land proclaiming "We still Love Lucy."[95] This sentiment also was expressed during the first filming of *I Love Lucy* occurring immediately after news of the scandal broke. Before the cameras began to roll, Arnaz explained to the studio audience that they may have heard some things about Lucy that were untrue. Arnaz told them, "Lucy has never been a communist. Not now, and never will be." Afterward, the audience stood up and applauded for over a full minute, with one audience member crying out, "We're with you!"[96] Arnaz concluded his speech by introducing Ball as "my favorite redhead. That's the only thing red about Lucy and even that is not legitimate." The day of the filming, California Representative and member of HUAC Donald L. Jackson came out in support of the actress, asserting that there was no real evidence she was party member.[97] Although Ball was eventually exonerated and was even invited to perform with the rest of the cast of *Lucy* for President Eisenhower only two months after the scandal, there was a small faction of the public who was not convinced of Ball's innocence. Employees of an Indianapolis car dealership tried to drive her off the air with a write-in campaign and a number of columnists wrote in favor of her dismissal.[98] One such editorial, published in the *Huntington Herald Dispatch*, contained this missive:

> We doubt if a single returning POW still haunted by Korea will give with the laughs the next time he is invited to "love" Lucy on the TV screen. The latest episode of Lucy and Desi has to be read and weighed against the background of the continuing expose of Communist infiltration of Hollywood, and not just as the disclosure of a fantastically naive incident in a top TV star's domestic life.[99]

However, by the end of 1953, the accusations against Ball were all but forgotten as accounts of her life with Arnaz and her two children once again took precedence in media coverage. The resilience of Ball and Arnaz's popularity in the face of such a potentially pernicious scandal was the final confirmation of the couple's influence over the American public. Their successful mobilization of the domestic, the proliferation of their images in syndication and merchandising, and the potency of their combination of their continuing characters and a situational narrative with vaudeville-inspired performance style became a blueprint for family comedies and stars during this period. In addition, their role as producers and primary managers of their images paved the way for other stars to negotiate new relationships with networks, packagers, sponsors, and talent agencies.

Stars and Independent Production

Although many classical radio and early vaudeo stars had negotiated director or producer titles into their contracts with networks and agencies, the role of the television star as producer was even more common and meaningful in the early era of independent telefilm production. Once again, Ball and Arnaz served as progenitors of this movement, as their positions as the heads of the influential Desilu company set the standard for the burgeoning relationships between networks and their talent. In addition, framing of their simultaneous roles as family-friendly stars and powerful businesspeople was mitigated by discourses on gender and domesticity.

As the press began to cover Ball's transition from film to television in 1951, the actress's new career became framed as one that was chosen to better foster her marriage and raise her infant daughter. Arnaz and Ball, who met in Hollywood in the late 1930s and were married in 1940, spent much of their relationship separated as Ball worked on her films and radio shows and Arnaz was often on the road with his orchestra. According to the couple, their forced absences from one another took a toll on their relationship and, as a result, they almost divorced in 1944. When CBS offered Ball her own television show in 1950, Ball was extremely interested in the project but would only agree to it if the network signed on her husband as her co-star. After the network declined, saying that the American public would never believe that a Cuban and an American glamour girl were a married couple, Arnaz and Ball took their vaudeville comedy act on the road. When their performances were received to critical and popular acclaim, William Paley finally acquiesced and eventually agreed to film the program from the couple's home base in Hollywood. With impressive foresight and the help of his agent Don Sharpe, Arnaz offered a cut in his and Ball's salaries in exchange for the network's and sponsor's agreement to do the show on film and to give the couple all rights to the programs after their initial airing.[100]

The couple spun their negotiations with CBS as a gallant effort on their part to preserve the normality of their family. A typical example of this is the story presented in a 1952 article entitled "Everybody Loves Lucy." In the article Katherine Albert writes, "It was then, out of mature desperation, that the idea for their television show was born. If they could work together, then, at last, they could live together. . . . Everyone told them they were crazy to give up the sure things that had earned them so much money for a risky TV venture. . . . The Arnazes went right ahead planning, sure of one thing at least: they would never be separated again."[101] In direct opposition to other discourses circulating about television work that told of the physical and mental hardships that the medium brought to vaudeo talent, Ball and Arnaz's filmed television series was presented as a healthy alternative to other types of careers in entertainment. This was because of the advent of telefilm, which promised to eliminate much of the tension and rehearsal time of live television work, and the framing of the program's narrative as one that would replicate the wholesomeness of American family life. This was similar to the rhetoric surrounding *Ozzie and Harriet,* one of the most popular suburban sitcoms of the mid-1950s. Tinky Weisblat discusses how the Nelsons' extratextual representations of their work on television as a "family business" or "cottage industry" neatly "enveloped the new medium in domesticity by suggesting that the producers of television were more interested in families than in finance—and thereby helped to legitimate the TV industry."[102]

A similar phenomenon surrounded the Arnaz-Ball coupling as their formidable reputations as influential industry insiders was presented to the public as simply the happy ending to a complicated love story. Nonetheless, some ruptures occurred in this seemingly pristine presentation. For example, Louella Parsons reported in June 1952 that Mark Rabwin, Ball's doctor, was so worried that the actress would break down from exhaustion that he insisted that she spend her weekends in the hospital.[103] In an interview with the actress, Parsons confronted her on the possible hardship of television production, inquiring, "Don't you admit [television] is harder [than film work]?" Ball replied: "Yes it is harder because it's like making a picture a week. But we love it."[104] In addition, by 1955 rumors of marital discord began to appear in the tabloids.[105] Such slippages were managed and contextualized within the star couple's continued struggle for financial and familial prosperity. If it wasn't for this framing, Ball's image as an industrious and commanding television producer could threaten to unravel her claims to normativity, even though, as many current social historians point out, women were entering the workforce in the 1950s at unprecedented rates.[106]

Mary Desjardins discusses how "the tensions between the Arnaz's on- and off-screen personas and between 1950s gender norms and the demands of work surface in the need to promote Lucy and Desi as married stars, business people, and as fictional characters, the Ricardos, who *do* ultimately support contemporary familial ideology."[107] Although it was clear that Desilu was co-owned by both Arnaz and Ball, the press accounts of their working relationship often posited Arnaz as the one who was really in control. Despite the fact that CBS was originally only interested in Ball and that much of the network's early negotiations with the couple were based on the appeal of Ball's persona, Arnaz's acquisition of his title as executive producer of *Lucy* and (eventually) other programs such as *Our Miss Brooks* and *December Bride* was what was most prominent in public discourse. Certainly Arnaz and Ball were not the first married star producers/actors in Hollywood, or in television for that matter; yet their prominence did make this type of arrangement more visible and promoteable in the early 1950s.

Radio, television, and film actors in general were some of the most prominent independent television producers during the early years of telefilm production. Bing Crosby, who established Bing Crosby Productions in 1945, was an early leader in television production, reaching the peak of his success in this area in the 1960s with filmed programs such as *Ben*

Fig. 6.10 The Nelson family. Library of American Broadcasting.

Fig. 6.11 Ozzie Nelson directs an episode of *Ozzie and Harriet*. Library of American Broadcasting.

Casey, The Bing Crosby Show, and *Hogan's Heroes.*[108] Dick Powell formed Four Star Productions with David Niven and Charles Boyer in 1952, which became one of the largest program suppliers by the late 1950s. Bob Hope Enterprises actually entered into full partnership with NBC in 1957. At the time, the network and the star agreed to co-produce five theatrical films and forty hours of television programs over five years.[109]

Ozzie and Harriet, which premiered in 1952 and was another "family production," was tightly controlled by the head of the Nelson family, Ozzie. The program itself was created, produced, directed, and written by Nelson who had signed a television contract with ABC in 1949 after a five-year run on radio. He negotiated the desires of sponsors, networks, and agencies, and Nina Liebman claimed, "While Nelson did employ other writers (including his younger brother Don), he approved all script outlines prior to their completion, did the final rewrite, and, of course, had ultimate power on the soundstage or in the cutting room."[110] Other domestic sitcoms of the mid to late 1950s, such as *Father Knows Best* (1954–1955, CBS; 1955–1958, NBC; 1958–1962, CBS). *My Three Sons* (1960–1965, ABC; 1965–1972, CBS). and *The Donna Reed Show* (1958–1966, ABC) although not completely controlled by the program's stars, were co-produced by the central actors (Robert Young, Fred Mac-Murray, and Donna Reed).

Christopher Anderson contends that the phenomenon of stars as television producers can be explained in part by these actor's visibility in the public and industrial domains: "Because of [the stars'] reputations in radio or movies and the backing of established talent agencies, they often were able to get series commitments from sponsors and networks based on a single pilot episode. And as a result of these commitments, they found it easier to arrange bank financing to cover production deficits and company expenses."[111] Supporting their talents' desires to create their own production companies, talent agencies became increasingly involved in this area during the early 1950s. William Morris was offering networks packaged series (produced by other people) that showcased the stars it had under contract. For this service, the agencies would receive a 10 percent packaging fee that was based on their entire production cost of the program rather than just their stars' salaries.[112] They would continue this practice, primarily focusing on live programming rather than telefilm and most often offering their talent ownership rights to the production package. By 1953 they were receiving packaging fees for programs such as *My Little Margie* (1952–1953, CBS; 1953–1955, NBC), *Ted Mack's Original*

Amateur Hour, Your Show of Shows, and *Racket Squad* (1950, syndicated; 1951–1953, CBS). MCA packaged programs in this manner for a time, but then began to see increased profit potential with telefilm. Although the Screen Actors Guild (SAG) forbade agents to produce films, it had not yet ruled on television production in 1949, and MCA wanted to use this opportunity to get in on the long-term profits offered by telefilm production. As a result, MCA had formed Revue Productions to produce a few television programs. By 1954 (after Ronald Reagan, president of SAG, granted MCA an official waiver in 1952 permitting the agency to produce an unlimited number of programs even if SAG eventually ruled against talent agents' involvement in TV production), Revue Productions was making more money from its television productions than MCA was from its representation work.[113] Even though none of them would come close to being as successful as MCA, other talent agencies also had been granted waivers from SAG to enter into television production, including Charlie Feldman's Famous Artists. *Variety* reported in 1953 that networks were dissatisfied with the control that such agencies had maintained over network programming:

> Talent agencies are becoming increasingly stronger in the television scheme because of their control of headliners. . . . The networks need headliners as much as they need giveaways and low-cost mysteries. As long as that need continues the talent agencies will have a degree control that precludes a disassociation from the commission men.[114]

The article goes on to claim that by 1953 approximately "$50,000,000 in commissions will be channeled into talent agency coffers."[115] Yet, that same year, it was reported that MCA and William Morris were having a tough time sustaining some of their programs ratings. *Variety* reasoned that "if the talent agency shows have come by hard times on the rating front, a major reason lies in the fact that there's so much concentration on maneuvering the program to permit maximum use of the agency talent as to dissipate the value and effectiveness of the show itself."[116] In a maneuver that revealed the advantages of both producing programs and representing talent simultaneously, MCA lobbied to change talent contracts in a way that would increase their programs' chances of survival. MCA told networks in early 1953 that it was no longer interested in thirteen-week contracts for its talent. It now wanted no less than thirty-nine-week contracts in order to avoid cancellations because of early reports of low ratings. [117] Throughout the 1950s, MCA's Lew Wasserman had been looking to get even more involved in the television business than he already was and in 1956 he found his opening—in the form of a friendship with the new head of NBC Bob Kinter. Soon, MCA was providing NBC with an exceptional amount of programming, even though the legality of the relationship was in question. In fact, in the spring of 1957, Wasserman had managed to get fourteen MCA-based (some were produced by the agency, while others were sold as packages to the network) prime-time programs on NBC. The network needed the talent that Wasserman provided and the agency found that its close connection to NBC won them more and more clients.[118] By the 1959–1960 season, MCA was making approximately $200,000 a year on each of their shows and had become the largest producer of television series in the United States.[119]

Because of the increasing investments of stars and agencies in television production, relationships amongst primary industry players were becoming all the more complicated. Still, the television star was entering into the latter half of the 1950s as a central and indispensable producer of image and narrative in the ever-expanding world of television and the factors that helped intensify the packaging and exploitation of stars in the mid-1950s continued late into the decade. Some of the issues raised during the immediate postfreeze era remained unresolved as the search for new faces, the high costs of talent, the use of Hollywood stars in television, and the negotiation of commercialism in television's narratives continued to be hotly debated topics well after 1955.

The desires of the broadcasting industry's seminal agents during the late 1940s and early 1950s formed and framed television stardom. Sponsors, networks, and advertising agencies helped shape personas, narratives, and program mise-en-scènes to better serve the project of new consumerist ideologies, whereas television networks helped mold such personas into salespeople for the medium and its perceived inherent aesthetic properties and conventions. From the vaudeo star to the sitcom star, the most popular performers of the period demonstrated the centrality of stardom in the project of television's development and wider dissemination.

EPILOGUE

The importance that the broadcast industry placed on stardom during television's early years of dissemination reveals its centrality to the larger economic, aesthetic, and cultural structuring of the medium. Powerful individuals in the industry considered the use of presold talent essential to the project of attracting national audiences and big-name sponsors as well as to the structure and value of developing generic conventions. Moreover, the acquisition of a stable of such talent could substantially increase a network's public renown, a lesson the television industry learned from studio-era Hollywood as well as from its own experiences with radio. Although television would not develop a highly regulated and centralized star system, it did produce its own standardized management practices and interplayer relationships with regard to its talent. Specifically, the television industry balanced the interests and needs of sponsors, advertising agencies, independent production companies, and networks with that of talent unions, agents, and individual performers. It also established its own measurement devices and publicity strategies to track and respond to a particular star's popularity with audiences.

Clearly, postwar television manufactured omnipresent, culturally resonant star personas that were highly influential in consumer and programming decisions as well in the negotiation of greater cultural ideologies. Stars differentiated programming while pushing the ideology of consumerism and representing specific cultural types. Furthermore, the influence and longevity of images belonging to people such as Jack Benny,

Milton Berle, and Lucille Ball underscore the long-term resiliency and effects that television talent has had on American culture.

As television entered the 1960s, the networks' initial moves to wrest program production out of the hands of sponsors were beginning to grow into a highly centralized system of vertical integration. Like Hollywood in the studio era, the networks eventually came to control all phases of the television industry from production to distribution and exhibition. Michele Hilmes has described this period in television history, which lasted from 1960 through 1980, as one that was defined by the *classical network system*.[1] Although many of the practices and relationships developed by the late 1950s would simply become more solidified in the following decades of this classical era, television's star system most likely went through significant alteration. More research needs to be done into this period as to better understand the ways in which stars were employed during years of cultural and industrial change.

There are a number of interesting points of entry for further research on the topic of broadcast stardom. For example, a television historian could investigate how broadcast stardom was changed by the increasing use of the magazine format of television advertising in the early 1960s and the accompanying rise of production companies that were partially owned by networks. The types of stars that began to arise out of the classical network system were not only comedy stars but also dramatic actors. It would be interesting to know the ways in which genre influences star production with regard to the dramatic television actor and how different their construction and reception was from that of the sitcom and vaudeo performers. Obviously, the socioeconomic milieu of American culture went through drastic changes in the 1960s and 1970s; this also contributed to intriguing differences between stars of these decades and stars of the immediate postwar period.

Studying the history of television's production of stars also may help us to comprehend the current media environment better. The intricate relationships that developed between film, television, and publishing during these years of massive conglomeration, along with technological developments such as cable television and the Internet, further complicated the processes and relationships described in this book. Although the roots of what we see now have come from the struggle to define stardom in the 1940s and 1950s, new developments are occurring in its industrial and cultural negotiation. For instance, how have our notions of television celebrity changed in recent years? For one thing, it would seem that stars of both television and film are no longer so willing to push product—whether it be for a network or an advertiser. In fact, most American stars are so reticent about endorsements that the only place they will do such ads and commercials is in Japan—where the payment is high and the chances of Western audiences seeing them is low. Yet, television stars still do work as commercial signs and industrial connectors as they promote their programs, allow the use of their image for merchandise, and represent the house style of their network or channel. The growing presence of reality television in prime time also raises potentially complicated questions regarding the changing nature of celebrity on television.

Regardless of what particular questions get asked by scholars, this book is meant to stimulate further research in this area. Broadcasting's star system resulted out of the marketing and narrative strategies of networks, the business imperative of advertising agencies and sponsors, the professional development and public relations tactics of performers and their agents and the cultivation of consumerist ideologies. Stars were vital to the construction and dissemination of broadcasting in the United States, and their particular roles and meanings need to be better understood.

NOTES

Introduction

1. Hal Humphrey, "TV is No Maker of Stars," *Los Angeles Mirror*, January 15, 1953: 60.
2. Ibid.
3. I conclude my analysis at about the historical moment that Christopher Anderson's study of the interpenetration of film and television begins. In his book, Anderson cites the fall of 1954 as a key transitional period in the relations between the two culture industries. In October of that year, Screen Gems (a subsidiary of Columbia Pictures) premiered *The Adventures of Rin Tin Tin* on ABC and *Father Knows Best* on CBS, making them the first major Hollywood studio to enter into television production. Later that month, Walt Disney premiered *Disneyland* and David O. Selznick's spectacular celebrating the seventy-fifth anniversary of the invention of the lightbulb, *Light's Diamond Jubilee*, was broadcast on all networks. These programs heralded a new age of television programming and industrial relations. Consequently, new forms of television stardom would begin to take shape. In addition, *Light's Diamond Jubilee* boasted the first television appearances of major film stars such as Lauren Bacall, David Niven, and Joseph Cotten. See Christopher Anderson, *Hollywood TV: The Studio System in the Fifties* (Austin: University of Texas Press, 1994) 3–5.

Chapter 1: Radio and the Saliency of a Broadcast Star System

1. Jennifer Hyland Wang discusses the complicated transition of daytime soap operas in "'The Case of the Radioactive Housewife': Relocating Radio in the Age of Television," in *Radio Reader: Essays in the Cultural History of Radio*, eds. Michele Hilmes and Jason Loviglio (New York: Routledge, 2002) 343–366.

2. "TV As Showbiz Hero," *Variety*, March 3, 1948.

3. Lynn Dumenil, *Modern Temper: American Culture and Society in the 1920s* (New York: Hill and Wang, 1995) 59.

4. For example, children's bedtime stories occasionally were broadcast.

5. Michele Hilmes notes that even in the 1930s, more than 40 percent of NBC's programming consisted of music. See *Only Connect: A Cultural History of Broadcasting in the United States* (Belmont, CA: Wadsworth, 2002) 93.

6. One exception to this is Eddie Cantor. He appeared on WJZ in 1921.

7. Arthur Frank Wertheim, *Radio Comedy* (New York: Oxford University Press, 1992) 4.

8. For more on the development of broadcast policy see Thomas Streeter, *Selling the Air: A Critique of the Policy of Commercial Broadcasting in the United States* (Chicago: University of Chicago Press, 1996).

9. Ibid. 101.

10. Edgar H. Felix, *Using Radio in Sales Promotion* (New York: McGraw-Hill, 1927) 103.

11. For an in-depth discussion of the Happiness Boys, see Susan Smulyan, *Selling Radio: The Commercialization of American Broadcasting 1920–1934* (Washington, DC: Smithsonian Institution Press, 1994) 97–99.

12. Felix 89–90.

13. Smulyan 117.

14. Michele Hilmes, *Radio Voices: American Broadcasting, 1922–1952* (Minneapolis: University of Minnesota Press, 1997) 83.

15. Another early example of a high-budget program from this transitional period is the City Service Concerts featuring Jessica Dragonette.

16. Allison McCracken, "God's Gift to Us Girls": Crooning, Gender and the Re-creation of American Popular Song, 1928–33," *American Music*, Winter 1999, 365–421.

17. Timothy T. Taylor, "Music and the Rise of Radio in the 1920s America: Technological Imperialism, Socialization, and the Transformation of Intimacy," *Historical Journal of Film, Radio and Television* 22, no. 4 (2002): 437.

18. Hilmes, *Radio Voices* 87–96.

19. Sydney Head, *Broadcasting in America: A Survey of Television and Radio* (Cambridge, MA: The Riverside Press, 1956) 140.

20. J. Fred MacDonald, *Don't Touch That Dial! Radio Programming in American Life 1920–1960* (Chicago: Nelson-Hall, 1979) 147. MacDonald points out that in 1929 one network had 33 percent of its programming produced by ad agencies, 28 percent by the network, 20 percent by sponsors, and 19 percent by special program builders. See MacDonald 32.

21. Local stations were producing some of their own programming.

22. Smulyan 83.

23. Christopher Sterling and John Kitross, *Stay Tuned: A Concise History of American Broadcasting* (Belmont, CA: Wadsworth Publishing Company, 1978) 156–57.

24. Ibid.

25. Some of these dramas were commercial; others were sustained by the network. Hilmes discusses the way in which the variety show and anthology represented the cultural debates over high and mass culture. See Hilmes, *Radio Voices* 183–229.

26. John E. Hutchens, "Mostly About a Comic Named Cantor," *New York Times*, August 31, 1941: X10.

27. See Wertheim 90–91. Some of these behaviors were described by the announcer in an effort to include the home audiences. Yet, because Cantor had also been the first to invite audience participation within the studio, there was ready laughter for the visual. The audience interaction courted by Cantor and utilized by most radio stars by the late 1930s was a continuation of vaudeville's presentational style and the intimate comedian-audience relationship. Initially, the J. Walter Thompson agency separated Cantor from his audience by forbidding members of the studio audience to applaud or laugh, as this was common practice in radio's early years. Cantor broke that wall by directly and physically engaging with the audience against the wishes of his producers. Broadcasters, who had initially assumed that home audiences would not want to hear laughter and applause in the studio, were surprised to find that studio laughter had the curious effect of increasing the home audience's pleasure. For more on this, see Smulyan 121.

28. Hadley Cantil and Gordon Allport, *The Psychology of Radio* (New York Harper and Brothers Publishers, 1935) 15.

29. Michael Mashon, "NBC, J. Walter Thompson and the Evolution of Prime-Time Television Programming and Sponsorship, 1946–1958," Unpublished dissertation, University of Maryland, 1995.

30. Mashon uses *The Chase and Sanborn Hour* (Cantor) and *The Fleishmann's Yeast Hour* (Rudy Vallee) as early examples of Thompson's strategy. Later examples include *Kraft Music Hall* (Bing Crosby), *The Rudy Vallee Show*, and, interestingly, *Lux Radio Theatre*, which had the director Cecil. B. DeMille as its host. See Mashon 44–48.

31. Ibid. 47–48.

32. For further discussion of this subject, see Michele Hilmes, *Hollywood and Broadcasting: From Radio to Cable* (Urbana: University of Illinois Press, 1990) 58–59.

33. From unidentified paper in Cantor's scrapbook. UCLA Eddie Cantor Collection, Box 34.

34. "25 Outstanders of '28-'29," *Variety*, January 8, 1930: 110.

35. Henry Jenkins, *What Made Pistachio Nuts? Early Sound Comedy and the Vaudeville Aesthetic* (New York: Columbia University Press, 1992) 172.

36. Albert F. McLean, *American Vaudeville as Ritual* (Louisville: University of Kentucky Press, 1965) 112.

37. Smulyan 120.

38. NBC had learned through its audience research on *The Goldbergs* that Gentile listeners were accepting of, and even enjoyed, references to Jewish life and shows with largely Jewish casts. Because of this, NBC executives and sponsors were willing to retain and exploit the ethnicity of vaudeville come-

dians on radio, especially if the larger portion of the audience continued to tune in. For more on this, see Smulyan 115–16.

39. John Seymour, "Mirth with a Mission," *Tower Radio,* April 1934: 16.

40. A. M. Sullivan, "Radio and Vaudeville Culture," *The Commonweal,* December 13, 1935: 176–78.

41. Eddie Cantor Collection. Box 34, Scrapbook. UCLA Special Collections.

42. Eddie Cantor Collection.

43. Robert Taylor, *Fred Allen, His Life and Wit* (New York: International Polygonics, 1989) 284–85. From memo found in one of Fred Allen's scrapbooks.

44. Tino Balio, *Grand Design: Hollywood as a Modern Business Enterprise, 1930–39* (Berkeley: University of California Press, 1995) 161–2.

45. Hope and Crosby road movies were a top box of the early 1940s. They would send these scripts to their writers for radio shows/appearances and have them come up with ad-libs to give a sense of spontaneity in their screen performances.

46. Myrt and Marge, *Variety Film Review,* June 3, 1937: 5.

47. Hilmes, *Hollywood and Broadcasting* 63–77.

48. Ibid.

49. See Hilmes on radio boycott in *Hollywood and Broadcasting* 63–77.

50. See Richard Dyer, "A Star is Born and the Construction of Authenticity," in *Stardom: Industry of Desire,* ed. Christine Gledhill (New York: Routledge, 1991) 132–140.

51. Cathy Klaprat, "The Star as Market Strategy: Betty Davis in Another Light," *The American Film Industry,* ed. Tino Balio (Madison: University of Wisconsin Press) 1985: 351-376. See Thomas Schatz, "'A Triumph of Bitchery': Warner Bros., Bette Davis and *Jezebel,*" *Wide Angle* 10, no. 1 (1988) for discussion of star as house style and the ensuing struggles that occurred as a result of the option contract.

52. The struggle over Hollywood stars between studios and exhibitors in the Radio Ban also bespeaks the cultural hierarchy of film fame over broadcast celebrity. A Hollywood star would never abandon his/her filmmaking stardom for a chance at radio (unless careers were seriously waning). Yet radio stars (and this would later be true for television stars—even in the 1990s) often were eager to try their talents out on the screen after making their fortunes on the air. Radio stars began to appear in films in significant numbers beginning in 1932. Paramount's first venture in this area was *The Big Broadcast* (1932). Later films such as *You Can't Cheat an Honest Man* (1938), a continuation of W.C. Fields and Charlie McCarthy's on-air conflicts, and *Love Thy Neighbor* (1940) and *It's In the Bag* (1945), which were filmic representations of Benny and Allen's feud, played directly with the narratives and personas created in radio.

53. MacDonald 32.

54. Felix 100.

55. The fact that the use of a star's image for another company's campaign required the inclusion of the sponsor's name, underscores this point.

56. Felix 174.

57. See Richard Dyer, *Stars* (London: BFI Publishing, 1979), and Richard deCordova, *Picture Personalities: The Emergence of the Star System in America* (Urbana: University of Illinois Press, 1990).

58. Mashon 43.

59. Wertheim 148.

60. Hilmes, *Radio Voices* 104–5.

61. *Report on Chain Broadcasting*: Commission order No. 37, Docket No. 5060 (Washington, DC: Federal Communications Commission) 17.

62. For more on this subject, see Wertheim 263–82.

63. Ibid. 277.

64. The War and Navy Departments cited Benny and Bergen for entertaining the troops.

65. December 7, 1948, NBC Archives, Miscellaneous file, no. 38, Box 115. University of Wisconsin Historical Society.

66. "NBC Stars get Tape Go-Ahead," *Variety*, January 26, 1948: 29.

67. Wertheim 330.

68. "American Tobacco Cancels Out Benny's $250,000 Promotion Coin," *Variety*, August 27, 1947: 25.

69. Wertheim 330.

70. "U.S. Rules Benny Must Pay," *L.A. Herald and Express*, January 3, 1949. Benny received $1,356,000 for his 60 percent share in Amusement. The IRS sued him in 1949, claiming that the capital gains provisions did not allow for the inclusion of "personal service." Benny's lawyers took the matter to the U.S. Supreme Court and won the case later that year.

71. Eugene Lyons, *David Sarnoff* (New York: Harper and Row, 1966) 286. Given this, it is interesting to note that in 1936 Paley went after three of NBC's top stars—Eddie Cantor, Al Jolson, and Major Bowes—because NBC had been beating CBS in the ratings for a few years before that. See Lewis J. Paper, *Empire: William S. Paley and the Making of CBS* (New York: St. Martin's Press, 1987).

72. "Benny Now on CBS One Yard Line," *Variety*, November 17, 1948: 23.

73. Paper claims that when Benny first spoke to Paley about moving to CBS, he told him, "I can bring the boys" (referring to the NBC's other top comedians). Paper 116.

74. Lyons.

Chapter 2: "A Marriage of Spectacle and Intimacy": Modeling the Ideal Television Performer

1. "TV's Fine but has a Long Way to Go Before Big Comics Join Up, Say Groucho, Fibber & Molly, Cantor," *Variety*, March 30, 1949. Also see J. Fred MacDonald, *Don't Touch That Dial! Radio Programming in American Life, 1920–1960* (Chicago: Nelson-Hall, 1979) 147.

2. "Exhibs Want Film-Tele Tie-Up," *Variety,* June 23, 1948: 1, 16.

3. Norman Blackburn, "But Names Will Never Hurt Me," *Variety,* January 5 1949: 105.

4. Frank Rose, *The Agency: William Morris and the Hidden History of Show Business* (New York: Harper, 1995) 124.

5. Rose writes that, "Talent agents were as eager to break into television as performers. . . . Strategically [MCA executives] Abe Lastfogel and Jules Stein both agreed with Bill Morris about the prospects for a vaudeville revival on television. The Morris office, with a client roster that included Amos 'n' Andy, Burns and Allen, Eddie Cantor, George Jessel, Al Jolson, and Ed Wynn, not to mention Milton Berle—clearly had the edge." See Rose 129.

6. "Tele: The Legiter's Bonanza: Golden Points Out Paradox," *Variety,* June 18, 1947: 32.

7. Bob Stahl, "Legiters, Not Filmers or Radioites Should Inherit TV, Sez Langner," *Variety,* April 28, 1948: 33.

8. "Recruits from Hollywood," *Time,* October 5, 1953: 80.

9. "Video Opening its Purse Strings to Lure Top Names," *Variety,* December 14, 1947: 27, 32.

10. For more on this, see Christine Becker's excellent dissertation, "An Industrial History of Established Hollywood Film Actors on Fifties Prime Time Television." Unpublished disssertation, University of Wisconsin-Madison, 2001.

11. "Tele Needs Show Biz Vets," *Variety,* July 28, 1948: 31.

12. Eddie Cantor, "Candor from Cantor," *Variety,* January 12, 1949: 32.

13. George Rosen, "TV Puts Radio Stars 'On Spot': Audiences Have a Greater Choice," *Variety,* November 2, 1949: 1.

14. Norman Blackburn, "But Names will Never Hurt Me," *Variety,* January 5, 1949: 105.

15. Ibid.

16. Bob Hope, "The Road to Video," *Variety,* July 9, 1947: 29.

17. Albert Stilman, "The Literary Symposium on Television," *Variety,* July 9, 1947: 32.

18. Arthur Frank Wertheim, *Radio Comedy* (New York: Oxford University Press, 1992) 386.

19. "Flock of Radio Comics May Take TV Plunge in Fall," *Variety,* June 7, 1950: 1.

20. Jack Gould, "Mr. Allen on Video," *New York Times,* October 1, 1950: 119. Gould was kinder a month earlier when he said that Allen "scored 100%" with his "low pressure school of comedy" in his NBC premiere. Yet he also noted that the show itself needed more inherent excitement and gusto. See Jack Gould, "Radio and TV in Review," *New York Times,* September 25, 1950: 42.

21. Robert C. Ruark, "Fred Allen at War with TV," *Hollywood Citizen-News,* June 24, 1952.

22. Robert Taylor, *Fred Allen: His Life and Wit* (NY: Little, Brown and Company, 1989) 291.

23. Joe McCarthy, "What Do you Think of Television, Mr. Allen?" *Life,* July 4, 1949: 69-71.

24. John Crosby, *Out of the Blue* (New York: Simon & Schuster, 1952) 33.

25. Jack Gould, "Bob Hope's Debut: Comedian Stars in Easter Revue on Television," *New York Times,* April 16, 1950: X13.

26. William Boddy, *Fifties Television: The Industry and Its Critics* (Urbana: University of Illinois Press, 1993) 80.

27. Carroll Nye, "Youngsters Told to Train for Television," *Los Angeles Times,* May 24, 1936: C10.

28. As in Golden Age radio, the regionalized nature of television during this period mimicked that of the regional vaudeville circuit. Berle, for example, frequently referred to his audience as a New York audience, mentioning local landmarks and playing up regional inside jokes. As his audience grew in terms of its geographic makeup, he would be forced to alter many of these practices.

29. See "Cantor Sees Changes Coming in TV, Warns Vs. Studio Audience Ogre," *Variety,* August 30, 1950: 2.

30. Leo Bogart, *The Age of Television: A Study of Viewing Habits and the Impact of Television on American Life* (New York: Frederick Ungar Publishing, 1956) 31.

31. Lee de Forest, *Television: Today and Tomorrow* (New York: Dial Press, 1942) 194.

32. William Eddy, *Television: The Eyes of Tomorrow* (New York: Prentice Hall, 1945) 274.

33. Thomas Hutchinson, *Here is Television: Your Window to the World* (New York: Hastings House, 1946) 141–2.

34. Michael Mashon, "NBC, J. Walter Thompson, and the Evolution of Prime-Time Television Programming and Sponsorship, 1946–1958." Unpublished dissertation, University of Maryland, 1995: 85.

35. "George Burns Votes for Studio Audience," *Television Magazine,* December 1950: 13.

36. Edward Stasheff, *The Television Program: Its Writing, Direction and Production* (NY: AA Wyn, 1951) 21.

37. *Admiral Broadway Revue* with Sid Caesar and Imogene Coca in 1949 was the first variety show broadcast from a large New York City theater.

38. Rob Stahl, "Talent Fears Large Theaters: Alarmed at Recent Buys," *Variety,* August 8, 1950: 73.

39. "Cantor Sees Changes Coming in TV, Warns vs. Studio Audience Orge," *Variety,* August 30, 1950: 2.

40. "Texaco Star Theatre," *Variety,* September 29, 1948: 46.

41. *Colgate Comedy Hour* was broadcast in a large theater. However, when Pat Weaver noticed that his performers were having a hard time playing such a large audience for laughs, he had his production crew "put up a big screen with a projector lens so that above Hope, or whoever, you could look up and you'd see his face up there telling his joke…it was a great way, as I used to put it, to marry spectacle and intimacy." From Judine Mayerle's interview with Weaver. Mayerle, "The Development of the Television Variety Show

as a Major Program Genre at the National Broadcasting Company: 1946–1956." Unpublished dissertation, Northwestern University, 1983: 217. Interestingly, Arthur Godfrey spoke in favor of banning studio audiences altogether, arguing that "their presence distracts the entertainers from focusing their attention on the cameras." See "Godfrey for TV Ban on Studio Audiences," *New York Times*, November 23, 1949: 26.

42. See Mayerle 75 for an extended discussion of this.

43. "Hour Glass," *Variety*, May 15, 1946: 34.

44. "Sobol's Blueprint for Vaudeo Acts," *Variety*, May 22, 1946: 35.

45. Ibid.

46. Standard Brands was already sponsoring Bergen's radio show, which was the main reason that he was chosen (and agreed) to be the host.

47. "Hour Glass," *Variety*, November 20, 1946: 42.

48. Mayerle 70.

49. Ibid. 72.

50. "RCA Profit in Quarter Rises 23 Pct.: Video Set Sales Important Factor," *Chicago Daily Tribune*, May 5, 1948: B7.

51. *Variety*'s review of *Texaco Star Theatre*'s premiere episode reiterates the programs strong connection to vaudeville: "The '*Texaco Star Theatre*' projects a new kind of show-business—a revival of the era that had its greatest showcasing at the Old Palace Theatre before the days of air conditioning. The Texaco show, in the eighth-floor studios of NBC, is given in an atmosphere that isn't conducive to a top theatre presentation policy. Although the glamour of the theatre is partially destroyed by the cameras in front of the stage, and the studio discomfort is evident by the battery of lights that sometimes produces a Fahrenheit resembling a steam room, the ability of top performers to entertain in even the strangest kind of atmosphere is undeniable." Joe Cohen, "Texaco Star Theatre," *Variety*, June 16, 1948: 30.

52. Edwin H. James, "Anything for a Laugh!" *Redbook*, January 1950: 51, 94–96, 100.

53. "Milton Berle: TV's Whirling Dervish," *Newsweek*, May 16, 1949: 56–58.

54. "The Child Wonder," *Time*, May 16, 1949: 70.

55. "How Milton Berle Got Into Television," *LA Examiner*, April 29, 1950: 21.

56. "Highlights: '48-'49 Show Management Review," *Variety*, July 27, 1949: 35.

57. From an NBC ad in *Variety*, May 17, 1950: 28–29.

58. "The Child Wonder," *Time* , May 16, 1949: 70.

59. Karen Adair, *Great Clowns of American Television* (Jefferson, NC: McFarland, 1988) 40.

60. "The Child Wonder," *Time* , May 16, 1949: 72.

61. "The Child Wonder," *Time* , May 16, 1949: 72.

62. Sylvester Weaver memo from October 19, 1950, referenced in Mayerle 201.

63. In 1950, Weaver produced a list of forty-two "Big Name Comics" in a memo to himself. At some point during that year, he had put checkmarks next to the nine out of the top ten individuals/acts because by then they had been contracted to NBC. They included: Cantor, Allen, Martin and Lewis,

Danny Thomas, Hope, Berle, Groucho Marks, Fibber McGee and Molly, and Sid Caesar. "Comics," Sylvester Weaver memos, Box 118, folder 52, 1950. NBC Archives, University of Wisconsin Historical Society.

64. Contracts were actually drawn up by the network that forbade contracted comedians from appearing more then every other week on the network.
65. Mayerle 208.
66. Motorola, Pet Milk, Norge, and Fridgidaire bought into the show. See NBC Press Release, "NBC Year-end Report," January 15, 1951, Library of Congress, Motion Picture, Broadcasting & Recorded Sound Division.
67. In his autobiography, Caesar recounts the reason given to him by an Admiral executive. He claims that the executive told him that during the show's run they went from selling five hundred sets a week to ten thousand. In response, Caesar asks, "So what's so bad?" And, the executive replies, "We faced a difficult decision. We have just so much money, and we had to make up our minds whether to put it into capital investment, or to keep putting it into the show. Honestly, we had to put the money into capital investment and retool—just to keep up with the orders." Caesar replies, "What you're telling me is that maybe for the first time in history a show is being cancelled because it's bringing in too much business? We're being dropped because we were too good?" The executive smiles and says, "I'm so glad you finally do understand, Mr. Caesar." See Sid Caesar, *Where Have I Been?* (New York: Crown Publishers, 1982) 89–90.
68. *Saturday Night Revue* also was a further experiment in Weaver's planned move to his magazine concept of advertising. He allowed individual advertisers to sponsor *SNR* in half-hour blocks. Weaver wrote in his biography that *SNR* was "the final crystallization of several plans based on two concepts. The first was that we had to find a way to let the average-budge advertisers participate in high-cost programming, so that the base of the network's television income could be broad. . . . The second concept was that of all-evening programming." See Pat Weaver, *The Best Seat in the House* (New York: Alfred A. Knopf, 1994) 172. For more on this, see also Mayerle 171–74.
69. "CBS Grab for TV Comics Continues," *Variety*, March 29, 1950:1. Paley's comet quote came from "Paley's Return," *Variety*, November 16, 1949.
70. "Network of the Stars: The National Broadcasting Co," *Variety*, November 19, 1947: 30–32. "NBC Television Choice of America's Most Popular Stars," *Variety*, October 18, 1950: 36–38.

Chapter 3 : Lessons from Uncle Miltie: Ethnic Masculinity and the Vaudeo Star

1. Joseph Epstein, "Have You Heard the One About the Jewish Comic?" *Wall Street Journal*, April 1, 2002: A12; David Zurawik, "Mr. Television," *Baltimore Sun*, March 29, 2002: 1E.
2. Franklin Foer, "Kaddish for Uncle Miltie: Milton Berle, Television's No. 1 Jew," *Salon.com*, posted on March 28, 2002, at 2:19 P.M.

3. Although much was made about Berle's "life-long" contract with NBC, some have said that it was either a concession or intended as a way for NBC to control Berle, who was originally contracted to the Texas Company. In his biography, Pat Weaver claimed that the contract "was nothing more than a token of appreciation for his contribution to the network's welfare. I had no authority to offer him an actual lifetime contract, but he never forgot my gesture." Pat Weaver, *The Best Seat in the House* (New York: Alfred A. Knopf, 1994) 172.

4. Postwar television as its relationship to the question of age is a rather interesting one in that television began, at one point, to be the place that stars of a certain age could go to when other mediums, particularly film, were through with them. This, however, had a major downside, as film stars were reluctant to appear on television lest it seem that they were being put out to pasture.

5. Arthur Frank Wertheim, "The Rise of Milton Berle," in *American History/ American Television*, ed. John O' Connor (New York: Frederick Ungar, 1983) 56.

6. Judith Butler, *Gender Trouble: Feminism and the Subversion of Identity* (New York: Routledge, 1992) 136.

7. For example, it was reported in 1949 that Berle was born on West 118th Street. Robert Sylvester, "The Strange Career of Milton Berle," *Saturday Evening Post*, March 19, 1949: 38.

8. Douglas Gilbert, *American Vaudeville: Its Life and Times* (New York: Dover, 1940) 287.

9. Michele Hilmes, *Radio Voices: American Broadcasting, 1922–1952* (Minneapolis: University of Minnesota Press, 1997) 89.

10. Typically vaudeville ethnic routines were performed by a member of the ethnic group being portrayed (Irish performers played a "Paddy" character, for example). Interestingly, Jewish performers began to do minstrel routines in the 1910s. Cantor and Al Jolson continued this tradition on television. For more on this history, see Michael Rogin, *Blackface, White Noise: Jewish Immigrants in the Hollywood Melting Pot* (Berkeley: University of California Press, 1996), and Michael Alexander, *Jazz Age Jews* (Princeton, NJ: Princeton University Press, 2001).

11. Steven Seidman, *Comedian Comedy: A Tradition in Hollywood Film* (Ann Arbor: UMI Research Press, 1981).

12. Frank Krutnik, "A Spanner in the Works? Genre, Narrative, and the Hollywood Comedian," in *Classical Hollywood Comedy*, ed. Kristine Brunovska and Henry Jenkins (New York: Routledge, 1995) 29, 36.

13. Berle was quoted as saying, "Some people say I'm in the show too much and some people say I'm not in it enough. They accuse me of interrupting the acts but there's a difference between interruption and integration. . . . To come from left field and interrupt an act without any thought given to how it will look would be dangerous. If there isn't any actual reason to come into

the act, I assure you I will never do it." "How Milton Berle Got into Television," *LA Examiner*, April 29, 1951: 21.

14. Barry Rubin, *Assimilation and Its Discontents* (New York: Times Books, 1995) 98.

15. In his book, Vincent Brook traces the Jewish character in American sitcoms. His chapter on Molly Goldberg is particularly relevant to this discussion. See *Something Ain't Kosher Here: The Rise of the "Jewish" Sitcom* (New Brunswick, NJ: Rutgers University Press, 2003).

16. Ibid.

17. Karen Adair, *The Great Clowns of American Television* (Jefferson, NC: McFarland, 1988) 77.

18. Jenkins notes that "[w]hile the question of ethnic identity was a key concern within urban areas where the immigrant masses struggled to define their place in American society, and therefore a prime field for the construction of jokes, it did not prove as amusing in the hinterlands, which either found the question irrelevant or too threatening to provide much humor." Henry Jenkins, "Shall We Make It for New York or for Distribution?: Eddie Cantor, Whoopee, and Regional Resistance to the Talkies," *Cinema Journal* 29, no. 3 (Spring 1990): 43.

19. Ibid. 46.

20. Even the relatively innocuous signs of urbanity also caused problems for the entertainers as critics complained that their "big-city values" were offensive to small-town, middle-class viewers. References to performers' urban sensibilities were used in this context to connect their geographic origins negatively with the use of blue and ethnic humor and sexual innuendoes. Consequently, the same signs used to place a performer in a particular cultural milieu also were seemingly suspect to and potentially offensive to many television critics during this period. An argument could be made that there were anti-Semitic undertones to this conflation. Leslie Fiedler argues that "the discovery in the Jews of a people essentially urban, essentially Europe-oriented, a ready-made image for what the American longs to or fears he is being forced to become." Fiedler, "Saul Bellow," in *Saul Bellow and the Critics*, ed. Irving Maline (New York: New York University Press, 1967) 2-3.

21. Arthur Hertzberg, *The Jews in America: Four Centuries of an Uneasy Encounter: A History* (New York: Columbia University Press, 1986) 304–20.

22. Edwin H. James, "Anything for a Laugh!" *Redbook*, January 1950: 94–96, 100.

23. Douglas Gomery, *Shared Pleasures: A History of Movie Presentation in the United States* (Madison: University of Wisconsin Press, 1992) 85.

24. Rubin, 68-69.

25. Ibid.

26. The Friar's Club can be seen as an extension of this narrative, albeit a more private one. Most of the preeminent vaudeo performers of the period were members and socialized regularly at its New York facilities.

27. Maurice Zolotow, "The Fiddler from Waukegan," *Cosmopolitan*, October 1947: 138.

28. Henry Popkin, "The Vanishing Jew of Our Popular Culture: The Little Man Who Is No Longer There," *Commentary*, October 1952: 50.
29. For American Jews, the Rosenberg case became a major test of loyalty. For anticommunists, who had grown more strident since 1948, it meant dissociation with Jewish communists and denial that the Rosenberg case had anything to do with anti-Semitism. Most Jews ran for cover during the Rosenbergs' trial, feeling not only that it was a major test of their loyalty but also that it might be turned into an orgy of anti-Semitism. At the deepest levels, I suspect, it triggered all sorts of reactions associated with the Holocaust fears that had never been acknowledged or dealt with. See Morton Horowitz, "Jews and McCarthyism: A View from the Bronx," in *Secret Agents: The Rosenberg Case, McCarthyism, and Fifties America*, ed. Marjorie Garber and Rebecca L. Walkowitz (New York: Routledge, 1995) 262.
30. See Marjorie Garber, "Jell-O," in Garber and Walkowitz, *Secret Agents* 11-21.
31. Ibid. 15.
32. Jeremy Gerard, "Milton Berle Browses at Home and the TV Audience Gets a Treat," *New York Times*, December 11, 1990: C15.
33. "Be Careful on the Air: On TV, the Risk of Offending Is Even Greater than on Radio or in the Movies," *Sponsor*, September 24, 1951: 36–37, 75–80.
34. Marjorie Garber, *Vested Interests: Cross-Dressing and Cultural Anxiety* (New York: Routledge, 1992) 224.
35. For more on this, see Daniel Boyarin, Daniel Itzkovitz, Ann Pellegrini, eds., *Queer Theory and the Jewish Question* (New York: Columbia University Press, 2003).
36. Daniel Boyarin, *Unheroic Conduct: The Rise of Heterosexuality and the Invention of the Jewish Man* (Berkeley: University of California Press, 1997) 4–5.
37. Daniel Boyarin, *Unheroic Conduct: The Rise of Heterosexuality and the Invention of the Jewish Man* (Berkeley: University of California Press, 1997): 4–5.
38. Garber, *Vested Interests* 233.
39. Sandra Berle, "My Son, Uncle Miltie," LA Examiner, May 19, 1952: 21.
40. Gladys Hall, "Everybody's Uncle Miltie," Radio/TV Mirror, June 1951: 80.
41. Dorothy Rader, "The Hard Life, the Strong Loves of a Very Funny Man," Parade, March 19, 1989: In addition to gossip about his extramarital affairs, there were many rumors about the very large size of Berle's penis. Perhaps this discourse can be interpreted as an attempt to assert a more conventional masculinity for Berle to thwart questions about his sexual preference.
42. Andrew Sarris, The American Cinema: Directors and Directions, 1929–1968 (New York: E.P. Dutton, 1968).
43. Frank Krutnik, "Sex and Slapstick: The Martin and Lewis Phenomenon," in Enfant Terrible!: Jerry Lewis in American Film, ed. Murray Pomerance (New York: New York University Press, 2002) 111.
44. Ed Sikov, Laughing Hysterically: American Screen Comedy of the 1950s (New York: Columbia University Press, 1994) 190.

45. For more on Benny as, what Alexander Doty calls, "a gay straight man," see Alexander Doty, Making Things Perfectly Queer: Interpreting Mass Culture (Minneapolis: University of Minnesota Press, 1993) 63–80.

46. Hilmes, Radio Voices 194.

47. Margaret T. McFadden, finds a similar dynamic in Hope and Crosby's road movies. See Cohan's article, "Queering the Deal: On the Road with Hope and Crosby," in Hollywood Comedians: The Film Reader, ed. Frank Krutnik (New York: Routledge, 2003) 155–66.

48. Ibid. "America's Boyfriend Who Can't Get a Date: Gender, Race, and the Cultural Work of the Jack Benny Program, 1932–1946," *Journal of American History* (June 1993): 126.

49. Denise Mann, "The Spectacularization of Everyday Life: Recycling Hollywood Stars and Fans in Early Television Variety Shows," in, Private Screenings: Television and the Female Consumer, eds. Lynn Spigel and Densie Mann (Minneapolis: University of Minnesota Press, 1992) 41–70.

50. Jerome Beatty, "Unhappy Fiddler," American Magazine, December 1944: 28–9, 142–3.

51. Mary Livingston, "By His Own Doing, He's the 'Most Maligned Man in the World,' Says Mary Livingston of Husband Jack Benny," CBS press release, February 7, 1949, "Publicity file" Jack Benny Collection, UCLA Television Archive.

52. Stanley Gordon, "The Rebellion of Jack Benny," Look, May 8, 1951.

53. Ibid.

54. George Lipsitz, "The Meaning of Memory: Family, Class, and Ethnicity in Early Network Television Programs," in Spigel and Mann, Private Screenings, 75.

Chapter 4: "TV is a Killer!": The Collapse of the Vaudeo Star and Television's Talent Crisis

1. "TV Growing Old Too Fast," *Variety*, July 6, 1949: 32.

2. "TV Viewers Want New Names: Getting Tired of Same Faces," *Variety*, January 26, 1949: 31.

3. Abel Green, "TV Material and Programs," *Variety*, April 5, 1950: 35.

4. "TV Programming Problem Covered at NAB Meeting," *Advertising Age*, April 24, 1950: 33.

5. George Rosen, "TV Stars Play Hospital Time: Illness Rampant from Overwork," *Variety*, December 24, 1952: 21, 31.

6. Ibid. 31.

7. Milton Berle, "TV Is a Killer!" *This Week Magazine*, February 6, 1955: 12–13.

8. Max Liebman, "TV is Such a Challenge," *Variety*, July 29, 1949: 40.

9. Radio, too, struggled with what to do about the disposability of gags. Orrin Dunlap wrote that "the jokes [on radio] must be fresh. If the listener tunes in on some repartee he has heard before, no matter where, on the stage, screen or radio, he skips along the dial to another wave." He went on to say that one

of the tricks of the trade was for radio comedy writers to put a new twist on an old gag by changing some elements. For more on this see: "Furiously Proceeds Radio's Gag Hunt," *New York Times*, August 26, 1934: SM12.

10. "A-B-C-D's of Television Comedy," *Variety*, December 13, 1950.

11. Bert Briller, "Bergen Frightened at Casualty Toll of Top Comedians in Television," *Variety*, January 31, 1951: 1, 55.

12. "Comics Battle Hour Video Shows; Seek Showdown for Next Season," *Variety*, October 24, 1951: 1, 72.

13. Nielson ratings from the summer of 1952 revealed that half-hour programs were less popular with audiences than hour programs. See "Audiences Like 'em King Size: Hour TV Shows Pull Ratings," *Variety*, September 17, 1952: 37.

14. "TV a Full-time Job, Doubling Too Tough, Performers Find Out," *Variety*, March 30, 1949: 1.

15. The press accounts of Gleason mirror those of film star Fatty Arbunckle in many ways, particularly with regard to the ramifications of excess.

16. "His New Salary Tops Anything in TV But . . . Gleason Drives Himself Harder," *Life*, January 24, 1955: 23–29.

17. Peter Martin, "I Call on Jackie Gleason," *Saturday Evening Post*, July 1957: 36–37, 56–58.

18. Sid Caesar and Richard Gehman, "What Psychoanalysis Did for Me," *Look*, October 2, 1956: 49–52.

19. The move to Los Angeles would also help foster the developing relationships between networks and independent producers.

20. Pat Weaver, *The Best Seat in the House* (New York: Alfred A. Knopf, 1994) 205.

21. George Rosen, "NBC-TV Comics Coast-Bound: Cable will Cue Shift of Talent," *Variety*, July 4, 1951: 43.

22. Those comics who stayed in New York as well as many who made the move to Los Angeles took winter vacations during the 1951–1952 season under "doctor's orders." This presented an additional problem for the television industry as TV producers had "already exhausted virtually all available talent." Berle took two weeks off and was replaced by Perry Como and Ken Murray (both of whom already had their own shows). Instead of looking for new talent, producers felt the need, because of pressure from their competitors, to use established names as substitutes. See "Star Vacation Subs Newest Tele Problem," *Variety*, February 1, 1951: 26.

23. Rosen, "NBC-TV Comics Coast-Bound" 43. This article neglects to mention that Ed Wynn's CBS program originated in Los Angeles from 1949 to 1950.

24. "Advertisers' TV Appraisal: High TV Talent Costs Due to Agency Failure to Try New Faces—Jordan," *Variety*, October 31, 1951: 23.

25. Ibid. 23.

26. George Rosen, "Video's All-Star Tug-O'-War: Over-Pricing as TV Industry Evil," *Variety*, February 27, 1952: 1, 36.

27. "Advertisers' TV Appraisal" 23.

28. The high cost of vaudeo talent affected more than just sponsors. Because it was extremely difficult for regular television cast members to moonlight on stage and their success on the small screen caused their base salaries to skyrocket, vaudeville was eventually deprived of its top talent. By the summer of 1950, vaudeville had entered a slump "worse than vaude has experienced since the depth of the depression." Ironically, although the vaudeville aesthetic was experiencing a revival through broadcasting, its original stage form was dying off. "Despite a paucity of vaudeville theatres, prices of talent are expected to zoom considerably because of the sharp competition of video shows," wrote *Variety* in 1950. "Agencies are expecting the return of a seller's market this fall because of the multitude of television varieties which will be in direct competition with vauders and cafes for standard and name turning." See "Video Now Vaude's Villain" *Variety*, July 12, 1950: 1.

29. "Estimated Weekly Network TV Program Costs," *Variety*, April 26, 1950: 33.

30. "The Texaco Talent Cost Story," *Advertising Age*, September 11, 1950: 76–77.

31. Abel Green, "Milton Berle's NBC 'Dream Deal'; Works 20 Years, Collects for 30," *Variety*, March 21, 1951: 1, 13.

32. Memo regarding "NBC Talent Commitments," August 31, 1953, Weaver Files, 1953, NBC Archives, The Wisconsin Historical Society.

33. Bob Stahl, "Dumont's Texaco Flirtation May Spark TV Cost Appraisals," *Variety*, May 23, 1951: 34, 41.

34. "Sponsors Examine Themselves," *Variety*, April 9, 1952: 37.

35. Ibid.

36. Guest stars were charging between $2,000 and $3,000 in late 1950. *Texaco* paid as much as $3,500 and *Stop the Music*, $5,000. See George Rosen, "Television—A 'Lost' Industry: Agencies Now Champion Radio," *Variety*, March 5, 1952: 1.

37. "Industrial Relations Activities: Television and Radio," *Monthly Labor Review*, January 1951: 54.

38. "Hollywood to go Steady with TV: More Stars on Video; B.O. Aid?" *Variety*, September 28, 1949: 1, 30.

39. Christine Becker, "An Industrial History of Established Hollywood Film Actors on Fifties Prime Time Television." Unpublished dissertation, University of Wisconsin-Madison, 2001.

40. "Recruits from Hollywood," *Time*, October 5, 1953: 80–81.

41. Ibid. 80.

42. Ibid.

43. "What ABC-Paramount Merger Means to Sponsors," *Sponsor*, June 4, 1951: 32–33, 79.

44. "They're Off at Hollywood and Vine," *Advertising Age*, September 6, 1954: 38.

45. Rosen, "Television—A 'Lost' Industry: Agencies Now Champion Radio," *Variety*, March 5, 1952: 1.

46. George Rosen, "Wanna Buy a Radio Star?: Can't Sell 'Em Webs Lament," *Variety*, May 28, 1952: 1, 38.

47. Ibid. 1.

48. "Agencies to Harness Talent: Put Accent on No Name Shows," *Variety*, May 13, 1949: 27.

49. "Sponsors Crying for New Comics but Demand 'Ready-Made' Ratings," *Variety*, December 17, 1952: 1, 63.

50. George Rosen, "CBS Out on a Godfrey Limb: Off-Color TV Stirs Affiliates," *Variety*, March 22, 1950: 29.

51. Ibid.

52. In 1952, indecent content on television was just one concern of government officials. "At the moment five main areas offer continuing problems: 1) plugs that pail; 2) beer and wine commercials; 3) indecency; 4) racial stereotypes; 5) violence." See Bert Briller, "TV Code has upped Necklines, but Long Plugs, Beer Blurbs, Violence, Race Carbons Still Pose Problems," *Variety*, December 31, 1952: 20, 27. Also see Lynn Spigel, *Make Room for TV: Television and the Family of Postwar America* (Chicago: University of Chicago Press, 1992) 147–49.

53. New York–produced live anthology dramas were also considered to appeal to only "urban tastes." For more on this, see William Boddy, *Fifties Television: The Industry and Its Critics* (Chicago: University of Illinois Press, 1993) 93–107.

54. O' Meara, Carroll. Television Program Production. New York: Ronald Press Co., 1995: 47.

55. "Be Careful on the Air: On TV, the Risk of Offending is Even Greater Than on Radio or in the Movies," *Sponsor*, September 24, 1951: 36–37, 76–80.

56. "Low State of Comedy Blamed on Censorship, Pressure Groups," *Variety*, February 25, 1953: 1.

57. "How to Keep Reds off the Air—Sanely," *Sponsor*, November 5, 1951: 87.

58. "A Matter of Taste—and Should Talent Decide What's Good?," *Advertising Age*, February 1, 1954: 42, 43.

59. Ibid.

60. George Rosen, "Appraising the Video Comics: Vets Solidify Tele's Standing," *Variety*, November 8, 1950: 1, 42.

61. "TV Freezes Coin on Hot Names: Stabilizing Pay on Guest Stars," *Variety*, November 15, 1950: 1, 69.

62. Mike Kaplan, "Tout Television as Talent Tutor; Tube Increasingly Eyes New Faces," *Variety*, May 16, 1951: 2, 18.

63. Weaver 227.

64. George Rosen, "Pricing of TV off the Market: Berle, Others in Sponsor Trouble," *Variety*, March 4, 1953: 1, 38.

65. George Rosen, "TV's 'What's Next?' Dilemma: Casualty List Presents Poser," *Variety*, February 14, 1951: 1, 55.

66. Bob Chandler, "Webs Amateur Talent Binge: TV Success Cues AM Carryover," *Variety*, July 16, 1950: 33, 92.

67. "NBC: Now Beat Columbia," *Variety*, March 16, 1949: 25.

68. "TV Tryout Theatre for Broadway," *Variety*, October 31, 1951: 1.
69. Ibid.
70. Sylvester [Pat] Weaver, "Memo: Comedy Development Plan," November 19, 1951. Weaver Files, Box 119, file 72, NBC Archives, The Wisconsin Historical Society.
71. *Variety*, December 31, 1952: 38.
72. Ibid.
73. "NBC-TV to Build Comics and Writers for Every Medium," *Variety*, October 19, 1955: 1.
74. See Judine Mayerle, "The Development of The Television Variety Show as a Major Program Genre at the National Broadcasting Company: 1946–1956." Unpublished dissertation, Northwestern University, 1983: 150. She cites Ernest Glucksman Papers, uncatalogued, Special Collections Library, University of Southern California, Los Angeles, California.
75. Ibid.
76. Jack Gould, "TV Stars on Skids? Industry Shows Concern over Sagging Ratings of Prominent Entertainers," *New York Times*, May 15, 1955: X11.
77. Ibid.
78. George Burns with David Fisher, *All My Best Friends* (New York: G.P. Putnam's Sons, 1989) 290.
79. "Texaco Star Theatre," *Variety*, September 24, 1952: 31.
80. "CBS Preps Situation TV Comedies as Answer to NBC's Star Line-Up," *Variety*, February 6, 1951: 1, 36.
81. Ibid. 1.
82. "Variations On a Theme by 'Lucy,'" *Variety*, October 15, 1952: 32.
83. Spigel 151.
84. George Rosen, "TV's Changing 'Best Bet' Line-Up: 'Top 10' Rosters Bare New Trends," *Variety*, June 4, 1952: 25, 38.

Chapter 5: Our Man Godfrey: Product Pitching and the Meaning of Authenticity

1. "No Pitch Stars to Have Dim Future, K&E's Davis Warns," *Advertising Age*, April 19, 1954: 40.
2. "George Burns Votes for Studio Audience," *Television Magazine*, December 1950: 13.
3. Jack Benny was the other sponsor favorite for the way in which he was able to brand his show and associate his image with a product.
4. "Letters," *Saturday Evening Post*, January 14, 1956: 4.
5. "Before You Junk Your Commercial: 1,000 Members of TV Critics Club Reveal Which Favorite Commercials Make them Purchase the Product," *Sponsor*, January 2, 1950: 58.
6. "Godfrey Paid $440,514.16 in 1948," *Los Angeles Times*, May 18, 1949.
7. For example, see Jack Wilson, "No Business Like His Business," *Look*, October 6, 1952: 45–50. Robert Metz states the 12 percent figure in his book *CBS: Reflections in a Bloodshot Eye* (New York: New American Library,

1975). Also, Willam Boddy notes that Godfrey's case proved that "the financial impact of a single broadcast performer could be enormous to [CBS]" in his chapter, "Building the World's Largest Advertising Medium: CBS and Television, 1940–1960," in *Hollywood in the Age of Television*, ed. Tino Balio (Cambridge, MA: Unwin Hyman, 1990) 63–90.

8. Saul Prett, "Fight Rages on Over Godfrey," *LA Examiner*, January 14, 1953.

9. Jack O'Brian, *Godfrey the Great: The Life Story of Arthur Godfrey* (New York: Cross Publications, 1951) 62.

10. "Radio's Godfrey: The Man with the Barefoot Voice," *Magazine Digest*, June 1950, 29–35.

11. This program only ran for three months in 1950.

12. O'Brian 62. See also "Godfrey makes History with 3 Shows, 1 Broadcast, 2 Video in First Ten," *New York Times*, January 27, 1949: 44.

13. "Arthur Godfrey is TV-Proof." 62–64.

14. Isabella Taves, "Why Women Love Arthur Godfrey," *McCall's*, October 1953: 47–49, 59, 62.

15. Joe McCarthy, "Our Man Godfrey," *The American Weekly*, March 1, 1953: 5.

16. Joe McCarthy, "Godfrey's Golden Touch," *Cosmopolitan*, December 1952: 63–69.

17. Ibid.

18. "Arthur Godfrey is TV-Proof," 62–64.

19. "Dr. Dichter Doesn't Think the Commercial is Hurt by Tenseness Built Up by a Show," *Advertising Age*, November 17, 1952: 58–59.

20. Bill Davidson, "Arthur Godfrey and His Fan Mail," *Collier's*, May 2, 1953: 12.

21. Ibid.

22. O'Brian 38.

23. Thomas Streeter, *Selling the Air: A Critique of the Policy of Commercial Broadcasting in the United States* (Chicago: University of Chicago Press, 1996) xii.

24. Fred MacDonald, *Don't Touch That Dial! Radio Programming in American Life, 1920–1960* (Chicago: Nelson-Hall, 1976) 96.

25. Allen Havig, *Fred Allen's Radio Comedy* (Philadelphia: Temple University Press, 1990) 175. Havig quotes Max Wylie, ed., *Best Broadcasts of 1939–40* (New York: Wittlesey House, 1940) 156–65.

26. Michele Hilmes, *Radio Voices: American Broadcasting, 1922–1952* (Minneapolis: University of Minnesota Press, 1997) 211.

27. Havig 156–57.

28. Godfrey performed on Allen's radio show for six weeks in 1942. "Arthur Godfrey: He Has Empathy," *Time*, February 27, 1950: 72.

29. Ibid.

30. McCarthy, "Godfrey's Golden Touch" 66.

31. Ira Knaster, "That Man Godfrey," *TV-Mirror*, March 1948: 50–53, 91–94.

32. Ibid. 65.

33. Taves 62.

34. McCarthy, "Godfrey's Golden Touch" 64.

35. Taves 49.

36. Paul Whiteman, "Whiteman Sees Pix, Legit, Radio as Tributaries for 'Ol Man River Tele," *Variety,* August 6, 1947: 24.

37. Richard Dyer, "*A Star is Born* and the Construction of Authenticity," in *Stardom: Industry of Desire,* ed. Christine Gledhill (New York: Routledge, 1991) 132–41.

38. Richard deCordova, *Picture Personalities: The Emergence of the Star System* (Urbana: University of Illinois Press, 1990).

39. William Eddy, *Television: The Eyes of Tomorrow* (New York: Prentice Hall, 1945) 283.

40. Thomas Hutchinson, *Here is Television: Your Window to the World* (New York: Hastings House, 1946) 142.

41. Eddie Cantor. "I Like It!" *Variety,* July 11 1951: 43.

42. Donald Curtis, "The Actor in Television," in *The Best Television Plays, 1950–51,* ed. William I. Kaufman (New York: Hastings House, 1952) 320. For more on this, see William Boddy, *Fifties Television: The Industry and Its Critics* (Chicago: University of Illinois Press, 1993) 80–85. Also, for more on liveness and the perception of television's "realness," see Rhona J. Berenstein, "Acting Live: TV Performance, Intimacy and Immediacy (1945–1955)," in *Reality Squared: Televisual Discourse on the Real,* ed. James Friedman (New Brunswick, NJ: Rutgers University Press, 2002) 25–49.

43. DeCordova 113.

44. Denise Mann, "The Spectacularization of Everyday Life: Recycling Hollywood Stars and Fans in Early Television Variety Shows," in *Private Screenings: Television and the Female Consumer,* ed. Lynn Spigel and Denise Mann (Minneapolis: University of Minnesota Press, 1992) 41–70.

45. Edward H. Weiss, "Why is Arthur Godfrey?" *Broadcasting, Telecasting,* May 24, 1954: 104.

46. Ibid.

47. Ibid.

48. Prett 2.

49. Ibid.

50. Ibid.

51. John Crosby, "Arthur Godfrey: It Doesn't Seem Like Old Times," *Collier's,* September 30, 1955: 27.

52. "Letters," *Collier's,* November 11, 1955: 24.

53. Ibid.

54. Crosby 27.

55. Elaine Tyler May, *Homeward Bound: American Families in the Cold War Era* (New York: Basic Books, 1988) 166.

56. Tinky Weisblat, "What Ozzie Did for a Living," *Velvet Light Trap* 33 (Spring 1994): 21.

57. Of course, this veneer was often damaged by stories of Arnaz's infidelities and the couple's domestic battles. For a further discussion of this, see Mary Desjardins's chapter, "Lucy and Desi: Sexuality, Ethnicity, and TV's First Family," *Television, History, and American Culture,* ed. Mary Beth Haralovich and Lauren Rabinovitz (Durham, NC: Duke University Press, 1999) 56–74.

58. Wilson 47.
59. David Halberstam, *The Fifties* (New York: Villard Books, 1993) 527.
60. "That 'Fired by Godfrey' Tag Serves Acts as Springboard to Bigtime," *Variety*, April 20, 1955: 59.

Chapter 6: For the Love of Lucy: Packaging the Sitcom Star

1. "Performer or Pitchman?" *Variety*, November 11, 1953: 31.
2. "TV's Overexposed Comics Running for Cover; Fewer Shows, Less Time," *Variety*, July 16, 1954: 1, 37.
3. Jess Oppenheimer with Gregg Oppenheimer, *Laughs, Luck . . . and Lucy: How I Came to Create the Most Popular Sitcom of All Time* (Syracuse, NY: Syracuse University Press, 1996) 192.
4. "How They are Rated and What They Cost," *Variety*, January 13, 1954: 31. This article cited *I Love Lucy's* talent costs to come in at $38,000 per half hour, which was one of the best deals on television as the talent cost per one thousand people came in at ninety-one cents per one thousand people, as compared with NBC's lower-ranked *Life of Riley* at $1.47 per one thousand people.
5. "'Strange Things Happening': '53 Scorecard on NBC vs. CBS," *Variety*, February 4, 1954: 25.
6. George Rosen, "NBC to Talent: Get Realistic: New Contracts Sifted Closely," *Variety*, March 18, 1953: 21, 36.
7. "NBC-TV All-Star Revue Fades in April; Talent Contracts a Poser," *Variety*, January 21, 1953: 29.
8. Florence Small, "TV Picks up it's Costliest Check," *Broadcasting, Telecasting*, September 20, 1954: 103–4.
9. Frank Rose, *The Agency: William Morris and The Hidden History of Show Business* (New York: HarperCollins, 1995) 184.
10. Ibid.
11. "Jessel Gets Talent-Producer Contract," *Broadcasting, Telecasting*, March 9, 1953: 72.
12. "ABC-TV Reaching for the Stars; Bolger Signed Eye 2 Dannys," *Variety*, March 25, 1953: 1.
13. "ABC-TV Quest for Major Stars Sparks Battle of Webs for Talent," *Variety*, April 1, 1953: 1.
14. Rose 186.
15. "ABC-TV's $1,000,000 Ad Campaign Bonanza for Dailies; Even Ferries," *Variety*, September 8, 1954: 23.
16. "Push TV as Star-Maker for Pix," *Variety*, February 17, 1954: 3, 18.
17. "More TV Stars for Theatres: Sullivan Newest Switch-Hitter," *Variety*, May 11, 1955: 3, 6.
18. Ibid. 3.
19. "NBC-TV to Build Comics and Writers for Every Medium," *Variety*, October 19, 1955: 1, 78.
20. "ABC-TV's Get 'em Young: Major Payoff in Kid Accent," *Variety*, March 23, 1955: 21, 38.

21. Leonard Traube, "TV Drama Prescription: M.D.: Heavy Accent on Scalpels, Cough," *Variety*, September 21, 1955: 33, 48.

22. George Rosen, "The Dearth of a Salesman: TV Far Behind on 'Front Men,'" *Variety*, January 25, 1956: 21.

23. Ibid. 37. Jane Gaines quotes C. L. Yearsley, "Great Values on 'Stunt' Publicity," *Moving Picture World* 37, no. 3 (July 1918): 334.

24. Charles Eckert, "The Carole Lombard in Macy's Window," in *Stardom: Industry of Desire*, ed. Christine Gledhill (New York: Routledge, 1991) 36.

25. Ibid.

26. Ibid. 87.

27. Mary Anne Doane, *The Desire to Desire: The Woman's Film of the 1940s* (Indianapolis: Indiana University Press, 1987) 33.

28. George Jessel, "What Good is a Rating if You Don't Sell Product?" *Variety*, January 1, 1954: 83.

29. "Steve Allen—One-Man Ad Agency," *Variety*, September 29, 1954: 23, 60.

30. "What Merchandising Aids are Available from TV Stations, From Networks," *Broadcasting, Telecasting*, August 31, 1953: 82–86.

31. Ibid. 84.

32. Hal Davis, "Merchandising . . . It's a Must," *Broadcasting, Telecasting*, July 27, 1953: 87.

33. Walter A. Scanlon, "TV Films Need Merchandising: Eight ways it Can be Done," *Broadcasting, Telecasting*, October 12, 1953: 98.

34. In 1955, NBC brought *Home* to Mexico City, *Tonight* from Ohio and *Comedy Hour* from Tijuana. Val Adams, "NBC Programs Plan More Tours," *New York Times*, February 1, 1955: 35.

35. "Top TV Shows May Tour in Fall: Dealers Putting Heat on Sponsors," *Variety*, May 27, 1953: 25, 54.

36. Jane Gaines, "From Elephants to Lux Soap: The Programming and 'Flow' of Early Motion Picture Exploitation," *Velvet Light Trap* 25 (Spring 1990): 39.

37. Ted Mack, "'Hit the Road, Bud,'" *Variety*, July 29, 1953: 50.

38. Ibid.

39. "500G TV Package with Name Talent in 17-City Swing," *Variety*, April 1, 1953: 1, 69.

40. Ibid.

41. "Road Shows for TV Planned by Firm," *Broadcasting, Telecasting*, March 30, 1953: 40.

42. Ibid. 69.

43. Jack Benny Papers, UCLA Film and Television Archives, Jack Benny collection, Box 91, folder 1, "Promotion," 1950.

44. "750, Mostly Femmes, Lured to Caesar Show by TV's 1st Show Train," *Variety*, May 10, 1955: 27, 46.

45. Lynn Spigel, *Make Room for TV: Television and the Family in Postwar America* (Chicago: University of Chicago Press, 1992) 168.

46. Nina Liebman, *Living Room Lectures: The 1950s Family in Film and Television* (Austin: University of Texas Press, 1995) 110.

47. Ozzie Nelson, *Ozzie* (Englewood Cliffs, NJ: Prentice Hall, 1973) 241–42.

48. Elaine Tyler May, *Homeward Bound: American Families in the Cold War Era* (New York: Basic Books, 1988) 181. For more about the history of consumer culture in the United States, see also Lizbeth Cohen's wonderful book, *A Consumers' Republic: The Politics of Mass Consumption in Postwar America* (New York: Knopf, 2003).

49. E.F. Seehafer and J.W. Laemmar, *Successful Radio and Television Advertising* (New York: McGraw-Hill, 1951) 132.

50. Edward H. Weiss, "Why is Arthur Godfrey?" *Broadcasting, Telecasting*, May 24: 104.

51. Ernest Havemann, "Girl with a Rubber Face: On TV Impish Imogene Coca Parodies Everybody, Herself Included," *Life*, February 5, 1951: 53. Havemann's take on female comedian's chimes with Kathleen Rowe's definition of the "unruly woman." For more on this, see Kathleen Rowe, *The Unruly Woman: Gender and the Genres of Laughter* (Austin: University of Texas Press, 1995).

52. Denise Mann, "The Spectacularization of Everyday Life: Recycling Hollywood Stars and Fans in Early Television Variety Shows," in *Private Screenings: Television and the Female Consumer*, ed. Lynn Spigel and Denise Mann (Minneapolis: University of Minnesota Press, 1992) 46–48.

53. Raye was a regular host of *All-Star Revue* before the title was changed to *The Martha Raye Show* in December 1953.

54. "Why I'm Through with Big TV Shows," *Sponsor*, May 2, 1955: 94.

55. Jack Gould, "TV's Top Comediennes," *New York Times Magazine*, December 27, 1953.

56. Havemann 53.

57. "The Private Life of Imogene Coca," *American Weekly*, March 2, 1952.

58. Debs Myers, "The Funniest Couple in America," *Cosmopolitan*, January 1951: 98.

59. Clara Beranger, "How Dumb is Gracie Allen?: The Lowdown on a Feminine Secret One Lone Husband Has Kept for Years!" *Liberty Magazine*, September 1, 1934: 15.

60. Gracie Allen as told to Jane Kesner Morris, "Gracie Allen's Own Story: Inside Me," *Woman's Home Companion*, March 1953: 122.

61. "I really don't act; I just live what we're doing. George calls this being a natural. He says that Jack Benny and I are the only two naturals he's ever known." Ibid. 116.

62. Ibid. 122.

63. Patricia Mellencamp, "Situation Comedy, Feminism, and Freud: Discourses of Gracie and Lucy," in *Star Texts: Image and Performance in Film and Television*, ed. Jeremy Butler (Detroit, MI: Wayne State University Press, 1991) 320.

64. Joyce Antler quotes Angoff's article "'The Goldbergs' and Jewish Humor," *Congress Weekly* 18 (March 5, 1951) 13, in "A Bond of Sisterhood: Ethel Rosenberg, Molly Goldberg and Radical Jewish Women of the 1950s," *Secret Agents: The Rosenberg Case, McCarthyism & 1950s America*, eds. Marjorie Garber and Rebecca L. Walkowitz (New York: Routledge) 1996: 200–201.

65. Morris Friedman, "The Real Molly Goldberg," *Commentary* 21 (April 1954) 364.

66. For much more detail on Molly Goldberg, see Vincent Brook, *Something Ain't Kosher Here: The Rise of the "Jewish" Sitcom* (New Brunswick, NJ: Rutgers University Press, 2003).

67. Antler 199–200.

68. Alexander Doty, "The Cabinet of Lucy Ricardo: Lucille Ball's Star Image," *Cinema Journal* 29, no. 4 (Summer 1990): 11.

69. "TV Reshapes Comic Pattern: Takes More than Standup Routine," *Variety*, June 10, 1953: 27, 57.

70. "Death of the Television Star: Get Yourself Good Vehicle," *Variety*, March 31, 1954: 1, 46.

71. Ibid.

72. Hal Humphrey, "Lucy May Go On Forever," *Los Angeles Mirror*, April 24, 1953: 39.

73. Harry S. Ackerman, "The Player Plus the Play," *Variety*, January 2, 1953.

74. This type of product awareness would become even more acute in the late 1950s in "realistic" domestic sitcoms such as *Father Knows Best* and *Leave it to Beaver*, in which an incredible amount of energy and money was spent, showcasing the suburban lifestyles and accouterments of its main characters.

75. For more on this, see Spigel 136–80.

76. Ibid. 165.

77. "TV Team," *Newsweek*, February 18, 1952.

78. Ball often was quoted in interviews as saying that she learned much of her physical humor from Buster Keaton, with whom she shared an office while working at MGM.

79. Undated reference from Oppenheimer and Oppenheimer 161–62.

80. Jack Gould, "Why Millions Love Lucy," *New York Times Magazine*, March 1, 1953: 10.

81. "Beauty into Buffoon," *Life*, February 18, 1952.

82. Leonore Silvian, "Laughing Lucille," *Look*, June 3, 1952: 77–78.

83. Jess Oppenheimer, "Lucy's Two Babies," *Look*, April 21, 1953: 14–15.

84. Bart Andrews, *Lucy & Ricky & Fred & Ethel: The Story of I Love Lucy* (New York: E.P. Dutton, 1976) 11.

85. Kathleen Brady claims that such merchandise brought a net profit of $500,000 to Desilu in 1953 alone.

86. Andrews 11.

87. "The Goldbergs," *Broadcasting, Telecasting*, April 19, 1954: 22.

88. Ad, *Variety*, February 22, 1956: 37.

89. Mary Beth Haralovich, "Sit-coms and Suburbs: Positioning the 1950s Homemaker," in *Private Screenings: Television and the Female Consumer*, eds. Lynn Spigel and Denise Mann (Minneapolis: University of Minnesota Press, 1992) 111–41.

90. Reruns were extremely profitable for the performers. Leo Kavner in "Rerun Payments: Actors Love 'em," *Broadcasting, Telecasting*, September 13, 1954, explains that the resulting contract between SAG, independent producers,

and the Alliance of TV Film Producers (July 1952) provided that, after two runs, the actor must be paid 50 percent of his/her initial salary for the third and fourth run, and 25 percent for the fifth and sixth runs.

91. "Jackson Says No Proof Lucille Ball was a Red: TV Star Explains Registration," *Los Angeles Daily News*, September 12, 1953: 3.

92. Ibid. Arnaz would assert his allegiance to the United States in subsequent press releases and articles. A particularly good example of this is an article written by the actor for *American Magazine* in February 1955 (p. 84), entitled "America Has Been Good to Me," wherein Arnaz describes in detail his disgust for the Cuban government and his eventual immigration to the United States. He writes, "More than most people, I cherish the protection of my loved ones which America offers. My children will grow up to be free individuals because free opportunity to me . . . My adopted country, as I said, changed me completely. It has given me such wonderful gifts as humbleness, tolerance, appreciation of all kinds of people, sureness that I can compete honestly, and the ability to shoulder responsibility."

93. Christopher Sterling and John Kitross, *Stay Tuned: A Concise History of American Broadcasting* (Belmont, CA: Wadsworth Publishing Company, 1978) 364–65.

94. "Telegrams Pour in, Saying: 'We Still Love Lucy,'" *Los Angeles Daily News*, September 14, 1953: 2.

95. Kathleen Brady, *Lucille: The Life of Lucille Ball* (New York: Hyperion, 1994) 219–20.

96. "TV Show Stays On: Fans Cheer; Red Prober Defends Star," *Hollywood Citizen-News*, September 12, 1953: 1.

97. Brady 220.

98. "I Love Lucy, But Not as Much," *Huntington Herald Dispatch*, undated release found in the clippings file of the Academy of Arts and Sciences.

99. Thomas Schatz notes that Arnaz had surmised that his only chance of becoming executive producer of the show was if it was done on film in Los Angeles, away from the watchful eye of the Biow agency (Philip Morris ad representative). Arnaz also believed film would allow him more creative control over his final product. See Shatz, "Desilu, I Love Lucy and the Rise of Network TV," in *Making Television: Authorship and the Production Process*, ed. Robert J. Thompson and Gary Burns (New York: Praeger, 1990) 117–35.

100. Katherine Albert, "Everybody Loves Lucy!" *Los Angeles Examiner*, April 6, 1952: 7–8.

101. Tinky Weisblat, "What Ozzie Did for a Living," *Velvet Light Trap* 33 (Spring 1994): 21.

102. "Louella O. Parsons in Hollywood: Lucille Ball and Desi Arnaz," *Los Angeles Examiner*, June 29, 1952.

103. Ibid.

104. Brad Shortell, "Does Desi Really Love Lucy?" *Confidential* 2, no. 6 (January 1955).

105. See Elaine Tyler May and Stephanie Coontz, *The Way We Never Were: American Families and the Nostalgia Trap* (New York: Basic Books, 1992).
106. Mary Desjardins, "Lucy and Desi: TV's First Family in the 1950s and the 1990s," in *Television, History, and American Culture*, ed. Mary Beth Haralovich and Lauren Rabinovitz (Durham, NC: Duke University Press, 1999).
107. Crosby also owned a number of other companies including Bing Crosby Enterprises; Bing Crosby, Inc.; Bing Crosby Investment Corporation; Bing's Things, Inc.; Decros (affiliated with Decca Records); and Crosby-Jayson Corporation. He also owned part of the Minute Maid Corporation for which he was spokesperson. See Donald Shepherd and Robert F. Slatzer, *Bing Crosby: The Hollow Man* (New York: St. Martin's Press, 1981) 186–92.
108. Thomas M. Pryor, "NBC, Bob Hope Join in Film Deal," *New York Times*, February 7, 1957: 21.
109. Liebman 51.
110. Christopher Anderson, *Hollywood TV: The Studio System in the 1950s* (Austin: University of Texas Press, 1994) 65.
111. Rose 192.
112. Ibid. 195.
113. "Talent Agencies Rule Roost: Control in TV Grows Stronger," *Variety*, September 2, 1953: 23, 43.
114. George Rosen, "TV's Half-Billion $ Talent Tap: Agents Alone to Get $50,000,00," *Variety*, July 27, 1955: 1.
115. "Trendex Ratings Show You Can't Sell Acts and Also Create Shows," *Variety*, November 25, 1953: 30.
116. "Stars Demand 39 VidPix Weeks," *Variety*, January 21, 1953: 27.
117. Connie Bruck, *When Hollywood Had a King* (New York: Random House, 2003) 130–35.
118. Ibid.

Epilogue

1. Michele Hilmes, *Only Connect: A Cultural History of Broadcasting in the United States* (Belmont, CA: Wadsworth, 2002) 194.

Index